THE ALZHEIMER'S PLAN
CARING FOR A FAMILY MEMBER

Dr. Gerald L. Kovacich
and
Dr. Vicki L. Anensen-McNealley

authorHOUSE

AuthorHouse™
1663 Liberty Drive
Bloomington, IN 47403
www.authorhouse.com
Phone: 1 (800) 839-8640

© 2019 Dr. Gerald L. Kovacich and Dr. Vicki L. Anensen-McNealley. All rights reserved.

No part of this book may be reproduced, stored in a retrieval system, or transmitted by any means without the written permission of the authors.

Published by AuthorHouse 07/08/2019

ISBN: 978-1-7283-1822-6 (sc)
ISBN: 978-1-7283-1821-9 (hc)
ISBN: 978-1-7283-1820-2 (e)

Library of Congress Control Number: 2019908955

Print information available on the last page.

This book is printed on acid-free paper.

Because of the dynamic nature of the Internet, any web addresses or links contained in this book may have changed since publication and may no longer be valid. The views expressed in this work are solely those of the author and do not necessarily reflect the views of the publisher, and the publisher hereby disclaims any responsibility for them.

Cover photo taken by Dr. Kovacich at sunset around San Juan Islands, WA, USA

CONTENTS

Dedication vii
Acknowledgements ix
Preface xi
Buddhist Mantra xv
Quotation xvii

SECTION 1
INTRODUCTION

Chapter 1-1: Alzheimer's and Related Diseases Defined 1
Chapter 1-2: Dementia—Alzheimer's History and
               Current Research 13

SECTION 2
PRE-DIAGNOSIS TO CONFIRMING THE
POSSIBILITY OF ALZHEIMER'S

Chapter 2-1: Before Diagnosis 19
Chapter 2-2: Diagnosis 29
Chapter 2-3: Diagnosed with Alzheimer's — Now What
               Do You Do? 35

SECTION 3
FINDING THE BEST FACILITY

Chapter 3-1: Various Types of Facilities 45
Chapter 3-2: Finding the Best Facility for Your Loved One 51
Chapter 3-3: Applications, Admissions, Screenings and
               Orientations 59

Chapter 3-4: Facility Staff 67
Chapter 3-5: Other Residents 75

SECTION 4
FROM THE BEGINNING TO THE END

Chapter 4-1: Legal Documents 83
Chapter 4-2: The Beginning of the End 93
Chapter 4-3: The Final Journey — Hospice 99
Chapter 4-4: When the End Comes 103

SECTION 5
LESSONS LEARNED AND THE FUTURE

Chapter 5-1: Summary of Alzheimer's 111
Chapter 5-2: The Future of Dementia—Alzheimer's Disease 117
Chapter 5-3: What You Can Do to Help 123
Final Quote 127

APPENDICES

Appendix 1: Alzheimer's Checklist Plan 131
Appendix 2: Stories of Dementia—Alzheimer's Sufferers 137
Appendix 3: Memorial to Helen M. Kovacich 147
Appendix 4: Thoughts of Grandma and Grandpa Filmore 151
Appendix 4: About the Authors 153

DEDICATION

From Dr. Kovacich:

This book is dedicated to my mother, Helen Marie Kinek-Kovacich: Thanks mom for teaching me about compassion, patience, love and charity.

To my family: I hope if I am stricken with this dreaded disease, you follow the advice given in this book.

This book is also dedicated to the professionals who work at the very special "Houses of Compassion:" assisted living, retirement homes, nursing facilities who have been one of the inspirations for this book and who dedicate their professional lives to making the last days of our elderly in these facilities as comfortable as possible while treating them with dignity and respect.

From Dr. Anensen-McNealley:

This book is dedicated to my late grandparents; without them in my life I would not have acquired the incredible patience it takes to love a person who has dementia nor would I have obtained the insight necessary to know what compassionate care really is.

It is also necessary to dedicate this book to the specialist care staff in the memory care units with whom we have been honored to know, work alongside, as well as the residents and their families. Caring for someone with a dementing illness can be rewarding and frustrating all in the turn of a moment. Thank you for opening your doors and hearts to this type of work, whether chosen or otherwise.

ACKNOWLEDGEMENTS

Writers should always thank those who help make their books possible and in this case, the following people:

- Sandy Nichol, editor and longtime dear friend for doing another awesome editing job with a totally screwed up manuscript.
- Deanna E. Richards for giving me (Kovacich) the time, space and quiet surroundings that a writer needs.
- Kelly Stadler (CNA, Med. Tech.) for her friendship and kindness to my mom.
- Shannon DelCiello (LPN, RCD, Healthcare Consultant) also for her kindness to my mom and having to often explain to me (Kovacich) the do's and don'ts of dealing with my mom.
- The staff at our publisher for bringing this book to fruition.

PREFACE

This book is based on the devastating effects dementia, and in particular, Alzheimer's disease, has on a person and their family and friends. In most cases, when someone forgets a person's name, a movie title, or any little mental error, we tend to attribute it to old age or a "senior moment." It's not until these senior moments become the norm before someone suggests Alzheimer's as the cause.

Helen Kovacich (Dr. Kovacich's mom) often said: "Of all the things I miss, I miss my mind the most." Then she would laugh. We don't know where she read or heard that saying, but ironically, after saying it over the decades, little did any of us know that it would come true.

When she was diagnosed with a form of dementia called Alzheimer's, I conducted online research to find out more about it and looked for books that would tell me what my family and I might expect from a non-clinical, practical, hands-on point-of-view, like how to deal with a loved one on a day-to-day basis. I didn't find any that met my needs on a personal level; there were no books explaining what to do if… However, I did find online sites discussing such things as signs and symptoms, which were helpful in a general way.

As I looked back over the years since she was diagnosed, my family and I found that we were learning "the hard way" by day-to-day experiences in dealing with her and seeing her continued decline. Also, in looking back we were able to match her actions with some online "signs and symptoms." Well, hindsight is always 20-20 as they say.

In April 2013, almost two years after she had been medically diagnosed (although I saw signs increasing over the years that she was getting older, since she was over 80 at the time; medically diagnosed at about age 89), I decided to keep a diary of the interactions with my mother and how the disease was affecting her over time. I also encountered other residents living with dementing illnesses, both at the assisted-living facility where my mother lived, and other places I visited.

Then it occurred to me to write this book with Dr. Anensen-McNealley, whom I had met at my mom's assisted living facility and who has had experience with dementia on a personal and professional level for decades. We decided such a book would be beneficial to others who may be dealing with a friend or relative who has now been diagnosed with some form of dementia and have no clue as to the initial signs or how to deal with it. It is very easy just to say, "Well, she is just getting old." or "Of course his memory is not as sharp as before. What do you expect at his age?" The fact is dementia is a real issue that deserves careful attention and intervention. Practical resources, like this book, aid a family member to not only care for someone living with Alzheimer's disease, but also wade through the care options and resources available.

Through it all, we have come to appreciate the tremendous work done by the professionals at those facilities who serve our elderly. The only facilities that seem to make the news are those where the residents are abused or ignored, giving all the other professional caregivers a black eye. It doesn't make the news when you find ones that are doing a great job. I guess only bad news sells.

This book is also written to highlight the many unsung professionals who go through tremendous daily pressures, and heartaches as they try to keep a positive attitude day in and day out for years as they deal with the "residents", many of whom have become friends, watching them decline in body and mind and, one-by-one, die.

They are truly angels of mercy as they care for the elderly and often grow to love them - by helping them bathe, clean up after their bowel movements - sometimes showing much more love than the elderly person's own family.

Hopefully, after reading this book, you and your family will have a better understanding of what to expect from your loved one or friend who has been diagnosed with the onset and progression of Alzheimer's disease and, most importantly, what you can do about it.

As your loved one gets older, look for early warning signs of Alzheimer's disease. Don't excuse memory lapses and unusual behavior as just old age.

Your loved one's intent may not be to deceive anyone, but they just think of it as "no big deal".

By going through this process with my mother, she made me a better human being. She inadvertently taught me to live my life focusing on being patient with others, showing compassion towards all sentient beings, and expressing charity to all. Ultimately she showed me to love all life forms.

The more her Alzheimer's took hold of her life and our lives, the more we learned about showing compassion, patience, love and charity; first to her, then to others in the assisted living community where she resided for a several years, and ultimately to everyone.

Alzheimer's and other dementias are daunting diseases that invoke fear in the young and old alike. With millions of people afflicted with the condition, and rates rising at an alarming rate, it seems time to question current best practice in caring for these individuals. While many families struggle to care for their loved ones at home, more find it too overwhelming to handle alone.

Long term care options are available, including nursing homes, assisted living facilities, residential care settings and adult day care.

When you finished this book, we hope that you will:

- Have a better appreciation of dementia and especially, the symptoms of Alzheimer's
- Learn more about Alzheimer's through your own research and talking to professionals in the field including your doctor, hospice professionals and other caregivers
- Establish a "What-If Plan" in the event your loved one or maybe you show early signs of this affliction
- Look for signs of dementia in your loved ones;
- Coordinate with your doctor and/or loved one's doctor and an attorney regarding what to do when diagnosed.
- Constantly show compassion, patience, love and charity to your loved one when stricken with this disease.

OM MANI PADME HUM[1]

---

[1] **14ᵗʰ Dalai Lama**
"It is very good to recite the mantra Om mani padme hum, but while you are doing it, you should be thinking on its meaning, for the meaning of the six syllables is great and vast... The first, Om [...] symbolizes the practitioner's impure body, speech, and mind; it also symbolizes the pure exalted body, speech, and mind of a Buddha[...]"
"The path is indicated by the next four syllables. Mani, meaning jewel, symbolizes the factors of method: (the) altruistic intention to become enlightened, compassion, and love.[...]"
"The two syllables, padme, meaning lotus, symbolize wisdom[...]"
"Purity must be achieved by an indivisible unity of method and wisdom, symbolized by the final syllable hum, which indicates indivisibility[...]"
"Thus the six syllables, om mani padme hum, mean that in dependence on the practice of a path which is an indivisible union of method and wisdom, you can transform your impure body, speech, and mind into the pure exalted body, speech, and mind of a Buddha[...]"
-From Wikipedia
NOTE: We believe Buddhism is a philosophy for Life; not a religion though others differ.

QUOTATION

...seems to have reached the age where life stops giving us things, and starts to take them away[2].

---

[2] From Indiana Jones 4: The Kingdom of the Crystal Skull movie,

SECTION 1

INTRODUCTION

This section is divided into two chapters:

Chapter 1-1: <u>Dementia and Related Diseases Defined</u>. This chapter discusses not only Alzheimer's, but also other types of dementia.

Chapter 1-2: <u>Dementia—Alzheimer's History and Current Research</u>. This chapter discusses the history of Dementia-Alzheimer's and what is being done to find a cure for it.

CHAPTER 1-1

ALZHEIMER'S AND RELATED DISEASES DEFINED

Is it more painful to forget, or to be forgotten?[3]
— Joyce Rachelle

INTRODUCTION

This chapter discusses not only Alzheimer's, but also other types of dementia.

The Disease

When one ponders a terminal disease, first thoughts go to cancer. This destructive overgrowth of cells robs a person of health as well as time. It is easy to spot a cancer patient – emaciated and pale, often hairless and eventually too tired to survive. People often speak of cancer patients as fighters – "Marie fought a hard battle with breast cancer" or "Jeff spent five years battling leukemia." These depictions of strength are in stark opposition to the physical attributes of the typical cancer patient. Then, once a person battles the horrible disease, most likely with chemotherapy and radiation, she/he becomes a "survivor." Annual Relay for Life events can be found in just about every county in the nation where survivors walk

---

[3] alzheimer's quotes

in support of cancer research while relatives and friends of those who lost their battle walk in memory and honor of what could have been.

"Janis"[4] is currently fighting her own battle with metastatic lung cancer. Once strong and feisty, this 70-year-old friend is now what can only describe as tiny, bald, and still fighting the good fight. But she has a weary and exhausted look in her eyes, despite the smile she puts on as if a television director is prompting her for the camera. This type of disease can be terminal for some, ending life in a rather medically-driven way and often much earlier than planned; at the same time, it can turn a fighter into a survivor given just the right mixture of chemotherapeutic agents and radiation therapy (or, alternatively, a positive outlook and healthy food).

Regardless of the terminal disease, it typically requires specific treatment and either a "win" or "lose" at the end of the game. Dementia is not like that; dementia is always terminal. No matter the treatment, regardless of the efforts, the fight, the battle, or the passion, the person with dementia always dies in the end. There are no survivors' walks for those afflicted with this brain-robbing disease. Rather we witness the slow change over time, from a vibrant parent, grandparent, or other loved one with passions and hobbies to a mildly confused and perhaps frustrated and disheveled acquaintance then finally a stranger who is unable to recognize her own family.

People with dementia's lifespan varies – the average time between diagnosis and death for a person with Alzheimer's is eight years, but some people can live up to 20 years with dementia[5].

No matter the specific dementing disease, the symptoms' core commonalities include:

- recent memory loss,
- language changes,
- personality changes,

---

[4] Names and some circumstances changed to protect the privacy of individuals cited in this book.
[5] verywellhealth.com

- delusions, and
- disorientation.

The differing diagnoses and unique symptoms set each type of dementia apart from another, but the core five symptoms, listed above, are common to all dementias. Some doctors cannot differentiate amongst the different types of dementia, and may simply label someone with Alzheimer's or the most basic of all - dementia.

> *Types of Dementia. Dementia is a general term for loss of memory and other mental abilities severe enough to interfere with daily life. It is caused by physical changes in the brain. Alzheimer's is the most common type of dementia, but there are many kinds*[6]

Recent memory loss involves items easily recalled by the average person, such as what was eaten for breakfast that day or a doctor's appointment that happened yesterday. People suffering from dementia simply cannot recall recent events. If it's early in the disease process, the person may attempt to cover up the fact that they cannot remember – they tell stories or confabulate to fill in the gaps; late in the disease the person is no longer able to do this. As time passes, the person moves back in time; they will view themselves as much younger or living during a different time in their lives. This becomes particularly difficult for family members and friends, as the person afflicted with dementia might forget even close loved ones and either view them as strangers or mistake them for other people.

Language changes are common in all types of dementia, and the severity and type really depend upon the area of the brain that is affected. Initially the person will forget the names of common items, places, and people. The person might begin repeating words over and over again, or repeat what others say (called echolalia). The person may have difficulty finding the right words, or mix two similar words up such as "salt" and "sugar." Other people may begin to cuss or use slang when, prior to dementia, they would never have done so. Some people begin speaking

---

[6] alz.org

"word salad," meaning the string of words coming out of their mouths does not make any sense to us, and yet seems to make perfect sense to them.

Personality changes occur with all dementing illnesses as well. A person who has alway been prim and proper may begin using foul language or say what is on their mind without realizing that their words may be hurtful or disrespectful. They may interrupt conversations and appear rude. Other people living with dementia may only enhance their past personalities with more vocal and physical demonstrations. For example, there was a man who had always been viewed as a "lady's man" – he was the town flirt and was quick to tell the grocery clerk or bank teller how beautiful she was. Once he eased into dementia and moved to an assisted living facility, he not only told the female residents there how good they looked, he would grab their bottoms or breasts and try to hold hands with just about anyone who would allow it. Perhaps inappropriate to anyone without dementia, this type of behavior may seem perfectly normal to the person with it.

Delusions are false beliefs and go hand-in-hand with just about any person living with dementia. Delusions allow a person with dementia to make sense of their utter confusion. It is not uncommon for a female resident in a memory care unit, for example, to believe that her husband dropped her off there while he's gone fishing, or a male resident may believe he is at the facility working as a maintenance supervisor. Whatever the delusion, it is helpful to avoid orienting the person to reality, since they are not capable of understanding; therefore, they may become confused, freeze, seem lost. It's best to just go along with them as they will usually forget what they said.

> *One woman in an assisted living facility talked about her friend across the street who asked her to come over for a visit. The woman laughed and said she would not go out in the snow and then laughed saying her friend had sprinkled salt from a salt shaker on the sidewalk to melt the ice and snow. There were no houses across the street and it was summer - no snow or ice to be seen.*

Disorientation occurs because the brain is permanently damaged and new memories cannot be stored. It is as if each new day...or hour... or minute comes without a history to support how the person got where they are. The person may not know where they are, or where their room is, or what to do with a fork or a comb. They might not know if it is day or night, and get their sleep schedule turned around. He might believe it's winter and layer up on clothes in the middle of a sultry summer day. Their disorientation may require you, as a family member or caregiver, to reintroduce yourself each time you see them because they're unsure of who you are or what you are doing. This can cause frustration and anxiety for the person living with dementia, as well as the person caring for her.

ALZHEIMER'S

The most common and well-known type of progressive dementia, Alzheimer's disease, has hallmarks that include *sundowning, wandering, collecting*, and *combativeness*. While a person with Alzheimer's disease may not demonstrate all of these behaviors, the person typically will have at least one or more, and the behaviors may come as the disease progresses, and then go as the person's brain continues to lose function.

Sundowning is when confusion and agitation worsens in the late afternoon or evening time. There are various theories on the reasoning behind this phenomena; the most common is that the person has been trying all day to hold in the fact that they cannot recall the most basic of events, and grows tired and weary from doing so, eventually leading to overt behaviors that demonstrate being overwhelmed. This experience is frustrating for the caregiver, whether that person is a family member or a paid professional. It is as if the person with Alzheimer's has "flipped a switch" and become a totally different person later in the day. There may be no way to console them; they may yell or hit out at you or experience paranoia or search endlessly for an item they cannot name. Sundowning can be frightening for the person experiencing it as well as for the caregiver.

While anti-anxiety medications may prove helpful at times to manage sundowning behaviors, often a consistent routine and a calm, safe environment serves the person best as they work their way through the agitation and eventually become exhausted.

Wandering is also a common feature of Alzheimer's disease. The person afflicted may appear to be searching for something – opening and closing cupboard and closet doors, entering and leaving rooms without apparent purpose, or simply walking the same path over and over again.

Environmental safety as well as rest and nourishment are key to support a wandering person. Ensure there are areas where safe wandering can take place. Remove any clutter and lock doors that may lead to dangerous outdoor areas. Entice rest with chairs along the person's "walking path." Be sure that finger foods are available (and beverages) so the person can keep up their strength and replace lost calories during these active times. Shoes should fit well to prevent falls as well as blisters.

Collecting is another hallmark of Alzheimer's. People with this disease may take items that do not belong to them – they are not stealing but rather this behavior appears to soothe the person. There may be no rhyme or reason to the items collected – just about anything is fair game. One person with Alzheimer's even collected furniture; during the night when everyone in the memory care unit slept, this gentleman systematically moved all of the dining room chairs to one empty corner of the building.

> *When someone has a strange habit, e.g. collecting, removing items, it is best not to confront them and ask them to stop doing it. They probably won't remember doing it and even if so, confrontation no matter how gentle often confuses the person, maybe even freezing up and becoming even more confused.*

Sometimes collectors are know by the name "hoarders". This is not to say hoarders have a form of Alzheimer's. However, this may be a possibility.

Rather than telling the person not to "steal" or attempt to take the items back, it is best to let the person collect and store the items, and once they are asleep or distracted, the caregiver can place the items back to their original places. This may become tiresome, as the person will likely collect these items over and over again. It is important to have items to collect, clearing the environment of these items will only prompt other behaviors.

Combativeness is probably one of the most difficult and frustrating features of Alzheimer's disease. Having someone you love hit out at you, or bite you, or push you away when you are only trying to help can be off-putting to say the least. Once the other symptoms are well understood, it is easy to see why a person can express combativeness. Your loved one might not know who you are or that you are trying to help. They may believe that whatever task you are trying to assist with has already been done. Leaving the person be and re-approaching later may be the best remedy for all involved – it can decrease your frustration while easing the person's anxiety. Assisting the person to do the task for themselves, rather than the caregiver doing the task, also minimizes combative behavior.

VASCULAR DEMENTIA

Another common form of dementia, vascular dementia used to be termed "multi-infarct" dementia and is the result of altered blood flow in the brain, most often from a stroke or multiple strokes. Not progressive in nature like the other types of dementia, vascular dementia may be mild or severe depending on the location and significance of injury to the brain combined with the number of strokes or impeded blood flow a person has incurred. Many experts refer to this as "vascular cognitive impairment"[7] rather than dementia, in order to capture the true nature of the condition. People at highest risk for developing this type of dementia are those with uncontrolled hypertension, high cholesterol, diabetes, or a history of transient ischemic attacks (TIAs or small strokes) or stroke.

To expand on this notion of "impeded blood flow" consider that the entire body, including the brain, is a network of blood vessels. Some of these vessels are large, while others are quite small. When there is a condition or disease that affects these blood vessels, as in heart disease for

---

[7] https://www.nia.hih.gov/health/vascular-dementia-and-vascular-cognitive-impairment-resource list

example, *it affects the entire blood system, not just the heart*. And so the small vessels in the brain are affected as well.

What may make this type of dementia worse is the fact that it often co-occurs with Alzheimer's and other forms of dementing diseases. And so there may not be clear-cut symptoms that outline vascular dementia versus another form, should the person have more than one form. Likewise, many people with vascular dementia also suffer from depression; depressive symptoms can make the symptoms of dementia worse.

Symptoms vary, since different parts of the brain may be affected by the impeded blood flow. Initially there may be language difficulties and disorientation and confusion. As blood flow continues to impede, you may see more gradual changes in the person's behavior that are considered hallmarks of vascular impairment:

- impaired judgment,
- Impaired planning abilities,
- short attention span,
- uncontrolled emotions, and
- decline in attention skills.

Impaired judgment affects the person's abilities to move through the day making even the most trivial decisions. For instance, a person who had suffered a stroke and had residual memory loss and was thought to have vascular dementia although it was never diagnosed. A licensed nurse, told her family and friends that she had "totally recovered" and returned to work as a medication nurse only to admit much later that she could no longer read or understand numbers.

It is unclear if the impaired judgment is the disease itself in this case, or a stubborn woman who fiercely desired to return to her previous life. Regardless, the person with vascular dementia may look normal; a close relative or friend will likely note the change in the person's judgment.

Making and carrying out plans and **short attention span** also go hand-in-hand. The person may be totally unable to figure out how to create and carry out the most simple of plans. Making dinner, for instance, may look like this:

> *A lady with vascular dementia has decided to make steak and salad for dinner. The steak is frozen; she places it in the microwave to thaw and eventually figures out how to defrost the meat. She takes the lettuce, carrots, radishes, and dressing from the refrigerator and sets them on the counter, but she is unsure what to do with them. When it is time for dinner, you find the steak still in the microwave, the salad items on the counter, and the lady watching TV in the other room, completely oblivious to the notion that she was even making dinner in the first place.*

Emotions are controlled in the brain, and when blood flow is altered it makes sense that emotions may also be affected. This results in what is termed "inappropriate emotions for the situation." A person with vascular dementia may laugh when others are crying, or get angry for no apparent reason. These emotional roller coasters can be difficult for the family and caregivers as well as the person experiencing the disease.

LEWY BODY DEMENTIA

Lewy Body Dementia (LBD) has unique hallmarks that set it apart from other dementias and include:

- sudden and fleeting confusion,
- vivid and well-formed visual hallucinations,
- Parkinsonism,
- fainting (and falls associated with that),
- nightmares, and
- malfunctions in the autonomic nervous system (fluctuations in body temperature, blood pressure, and pulse with no apparent reason).

Sudden and fleeting confusion may happen early in the disease, A person may seem cognitively intact, then suddenly become profoundly confused. This may last minutes, hours, or days. And then, without warning, the confusion remedies and the person is able to hold a conversation and recall memories. There appears to be no rhyme or reason to the confusion – no specific times of day or stressors that may prompt its arrival and departure. Eventually, as the dementia progresses, these fluctuations become less and less often.

Hallucinations in a person with LBD will likely have – most often these are visual but can also be auditory or touch. For some reason yet undiscovered, the most common hallucinations include those of animals, children, and dead relatives/loved ones. These hallucinations are vivid and well-formed; the person can describe, for example, the color of the [not really there] dog sitting at his feet. He can tell you the breed. He may talk to the dog, reach down to pet it. The dog may interrupt a conversation the person is having with you. Hallucinations are common in LBD; most of these are visual, but they can also be auditory or touch.

> *A gentlemen in an assisted living facility complained to the nurse in the middle of the night that he couldn't sleep as there were children playing and making noise in the hallways outside his room. The nurse responded and of course didn't find any one in the hallway. The nurse told the resident that they were gone. He calmly went back to sleep.*

Parkinsonism occurs with LBD, since the proteins in the brain of both LBD and Parkinson's disease are similar. The person's face may no longer show emotion, and they will likely begin shuffling when they walk. Stiffness in movements and joints also make walking and doing common tasks difficult.

Fainting may cause a person, without warning, to lose consciousness and fall. These falls are difficult to prevent, since it is the disease itself that causes them. Serious injury is likely, as the person is unable to "break the fall" by grabbing an object nearby or sliding to the floor. Likewise, the autonomic nervous system misfires, causing high and low body temperatures absent infection, as well as wide fluctuations in blood pressure and pulse. The person may experience profuse sweating as well as constipation and frequency or urgency of urination.

Nightmares are frequent and acted out – for example, if the person dreams of being chased, he may actually move his arms and legs as if he were running; they may also yell. Upon awakening, the person will have a difficult time differentiating between those dreams and reality; this may result in the person accusing others of stealing or beating him up or whatever the topic of the nightmare may have been.

Medications typically used to manage or minimize behaviors and symptoms in people with dementia may not work on people with LBD – these medications may increase the Parkinsonism and/or decrease the person's cognition. Anti-anxiety medications tend to have the opposite effect on people with LBD; rather than calming the person down they may increase their anxiety and aggression.[8]

FRONTOTEMPORAL DEMENTIA

Frontotemporal Dementia (FTD) has several aspects that sets it apart from Alzheimer's disease and other dementias. The protein accumulation and death of brain cells in this form of dementia tend to focus on the frontal and temporal lobes of the brain affecting:

- speech,
- mood/personality changes, and
- coordination.

Perhaps most unique, FTD typically hits between the ages of 40 – 60, setting it apart from other dementias. Another unique feature is the notion that memory loss tends to occur in the latter part of this progressive disease, rather than serving as the initial symptom.

People affected with FTD are younger, and so their loved ones (often a spouse or children) or co-workers may note a significant change in the person's mood and personality. Most people don't connect these types

---

[8] https://lewybodyresourcecenter.org/what-is-lbd/treatment-important-information/medications/

of changes to dementia but rather a "mid-life crisis" or depression and therefore result in marital problems and/or divorce and job loss.

Speech changes initially, and may simply become less or absent (aphasia). Other changes may include the person's speech becoming more hesitant or halted, which may appear to sound rude or abrupt. The person loses his social filter, and may say things that are distasteful or inappropriate and hurt others' feelings.

Mood swings may be experienced by people with this form of dementia. The person with FTD also has **personality changes** – they may become very egocentric, unable to acknowledge or empathize with others. Self-gratification becomes central to the person's life and may result in socially inappropriate behaviors like masturbating in public, overeating, or binge drinking.

Coordination is also affected, caused by muscle rigidity or weakness. This may lead to falls or the need for assistive devices such as walkers or wheelchairs.

SUMMARY

Some researchers surmise there are upwards of 60+ different diseases that include dementia as a symptom. While this book includes the most common types, there are resources (most available on the Internet) to discover more about those types not mentioned in this chapter.

Regardless of the type of dementia your loved one may have, there are commonalities that all people living with dementia share: memory loss. Perhaps the details of each disease miss some significant emotions that both the person living with dementia and the loved ones share – frustration, fear, and perhaps anger. And yet through all of this, love, compassion and understanding, patience and kindness can ease those emotions.

CHAPTER 1-2

DEMENTIA—ALZHEIMER'S HISTORY AND CURRENT RESEARCH

"We remember their love when they can no longer remember."
— Unknown[9]

INTRODUCTION

This chapter discusses the history of Dementia-Alzheimer's and what is being done to find a cure for this deadly disease.

History

The history of Alzheimer's disease and other dementias could easily be placed on a linear timeline graph, from the first diagnosis in 1906 to the creation of the National Institute on Aging in 1974 to the first medication approved by the FDA to treat Alzheimer's in 1993 to Barack Obama's signing of the National Alzheimer's Project Act in 2011.[10]

Yet a timeline really does not tell much about the history of the disease itself. It serves, rather, to call out the fact that many dedicated people have worked to combat the disease, whether through research or raising

---

[9] alzheimers.net
[10] https://www.alzheimers.net/2013-12-30/history-of-alzheimers/).

awareness of its existence. When did dementia start? How long has it been around? Why does it exist at all?

Alzheimer's disease was first described by psychiatrist Dr. Alois Alzheimer in 1906[11]. The patient's symptoms included memory loss and paranoia. The brain changes could not be truly evaluated until after the patient died. The autopsy showed "plaques and tangles" as well as significant shrinkage of the entire brain, both revelations that have become hallmarks of Alzheimer's disease.

Of note, this type of dementia as researched and reported by Dr. Alzheimer encompassed one woman in her early 50's who expressed symptoms similar to what we now recognize as Alzheimer's disease. At no time, at least as documented, did Dr. Alzheimer seem enthralled with what has become known to us today as Alzheimer's Disease - mainly because the fact that it is much more common in the elderly. He and subsequent researchers seemed to have felt that "senility" in the older population was simply part of the aging process, and not a disease at all. Although there are certainly more and more cases of early-onset Alzheimer's today than in past decades, the disease is much more prevalent in the elderly.

For much of the 20th century all dementia-related diseases were deemed to be Alzheimer's Disease. In 1984 the National Institute on Aging (NIA) began funding research and treatment for dementing illnesses; this research resulted in the first drug approved in 1993 to treat Alzheimer's. Ten additional drugs would eventually be approved over the course of the next ten years and a handful of these drugs are used today.

None of these medications, however, cure the disease. If caught early enough, the start of one or more of these medications may slow the progress of the eventual forgetting. The medications offer false hope to a still-terminal patient and their family.

As research has evolved, different types of dementia have been discovered. While there are around at least 50 different diseases that include dementia as a primary feature, a few stand out as the most commonly seen.

Lewy Body Dementia (LBD) is named after the neurologist Friederich Lewy, who in 1912 discovered unique proteins in the brains of people

---

[11] https://www.alzheimers.net/alois-alzheimers-21216461

suffering from dementia-like illness[12]. Lewy Body Dementia is thought to be the second most prevalent type of dementia, behind Alzheimer's, although many experts hold the second most common form as vascular dementia.

Frontotemporal Dementia encompasses various different dementing conditions that affect the temporal and frontal lobes of the brain. Historically, this disease was first identified in the early 1900's by psychiatrist Dr. Arnold Pick. Of note, Pick identified unique proteins in the brain called "tau proteins" that serve as the hallmark identifier of the disease. Research showed this particular form of dementia as affecting a much younger population, with victims, upon diagnosis, ranging in age from 45 – 65.[13]

Current research on dementia-related diseases is far-reaching and prevalent. Much research, of course, includes searching for new medication to slow or stop the progression of the symptoms. But much is being done in the way of hypothesizing the causes of the diseases themselves. It seems that new studies are being published every week, outlining possible causes and the need for additional research.

One cannot ignore the fact that dementia-related diseases were almost unheard of 200 years ago. While it makes sense that the lifespan has increased over that amount of time, it is also true that dementing conditions are becoming more the norm than not as people enter their elder years. There are still places in the world where the symptoms of dementia as we know them simply do not exist or are so incredibly rare that they are considered an anomoly. Specific areas in Japan, Costa Rica, Italy, France, Greece, Spain, and Central and South Americas all enjoy very little Alzheimer's disease.[14]

Some of the major links that connect increasing numbers of people with dementia encompass one's lifestyle. Obesity rates are skyrocketing throughout industrialized countries; this phenomena is a compilation of factors including lack of exercise and a diet low in vegetables and fruits and

---

[12] https://www.lewybody.org/about-dlb/science/
[13] http://www.whonamedit.com/doctor.cfm/1100.html
[14] https://www.awakingingfromalzheimers.com/coutries-with-the-lowest-dementia-rates-follow-these-7-food-secrets/

high in processed food-like items and sugar. These two major issues – lack of exercise and poor diet – lead to heart disease and diabetes.

Heart disease and diabetes cause changes in the blood vessels of the body...including the brain. Insulin resistance, which is a factor leading to diabetes, also affects the brain; Alzheimer's has become known as Type 3 Diabetes, to express the huge impact that the effects of circulation and insulin resistance may have on the overall operations of the brain.

Another theory and current research, which may connect as well to other chronic illnesses throughout the body, is the notion that inflammation plays a role in the development of Alzheimer's Disease and perhaps other dementias as well.[15]

SUMMARY

The future of dementia is a bleak one. Researchers anticipate a drastic increase in numbers of older adults who will live with at least one form of dementia. Currently one in ten adults age 65 and older have Alzheimer's Disease. Projections pose significant concern in the health care arena, with worry regarding how to pay for the care of these individuals, and just who will provide the care.

---

[15] https://www.alz.org/facts/

SECTION 2

PRE-DIAGNOSIS TO CONFIRMING THE POSSIBILITY OF ALZHEIMER'S

This section is divided into three chapters:

Chapter 2-1: <u>Before Diagnosis</u>: Seeing the changes, discusses a person's gradual mental decline that goes from normal to "old age" indicators to indicators of Alzheimer's. It also discusses the differences.

Chapter 2-2: <u>Diagnosis:</u> about sufficient Alzheimer's indicators that leads a family member to take the relative to a doctor to determine the possibility of Alzheimer's or other forms of dementia.

Chapter 2-3: <u>Diagnosed with Alzheimer's - Now What Do You Do?</u> This chapter discusses what to do with your loved one once diagnosed with Alzheimer's.

CHAPTER 2-1

BEFORE DIAGNOSIS

Memory is the diary that we all carry about with us[16].
— Oscar Wilde

INTRODUCTION

radual mental decline can go undetected for quite some time, particularly with a loved one who is seen every day. Subtle changes are easy to miss or view as a normal part of aging, There are distinct indicator's of Alzheimer's disease that, when known, can help guide you through the process of identifying whether symptoms are normal or something is far worse.

Getting Old

As your loved one gets old, it is important that you begin to look for signs of Alzheimer's Curse as early detection and subsequent action helps to ensure that they get the aid and support needed. While there isn't a cure right now, proper care and treatment may help delay a bad situation from getting worse by slowing down this disease's effects.

When your loved one begins forgetting, is that "just old age" or something more serious such as Alzheimer's, which makes up about 60-80% of dementia sufferers?[17] Often, when someone can't recall a name of a person, where they put their keys and the like, you as the relative of that

---

[16] brainyquote.com
[17] alz.org

loved one may tend to think of it as they are just getting old. After all, we all tend to forget something on occasion, such as names of people we have been introduced to.

Today, people tend to be living longer and as they do, their body parts, like an old car, begin to break down more often. It's just natural and in the past, such forgetfulness was rationalized as "just old age". However, over the last several decades, there has been a growing awareness of Alzheimer's disease.

Let's begin by discussing "Before Diagnosis". According to various medical professionals, Alzheimer's disease strikes at all levels; however, the majority are over 65 years old.[18]

Take 80 year old Clara as an example (A fictitious person who meets some of the general criteria of Alzheimer's sufferers.). Like so many women her age, she is a widow, lives alone, but has relatives and friends close by. She has always been independent and prides herself in being able to take care of herself. When talking to her friends and family members, she sometimes complains that she seems to be getting more forgetful as the years go by. Her friends and relatives think little of it and say, "Oh you're just getting old." But is that the only reason? She replies, "Yeah, getting old is hell!" They all laugh about it and move on in the conversation.

> *Because African-Americans are more likely to have vascular disease (problems with blood circulation), they may also be at greater risk for developing Alzheimer's. Risk factors for vascular disease — like diabetes, high blood pressure and high cholesterol — may also be risk factors for Alzheimer's and vascular dementia*[19]

As time passes by, some of the relatives or friends continue to think nothing of it, avoiding any thoughts of her developing Alzheimer's. On the other hand, others are thinking that maybe she is getting Alzheimer's. However, few say anything about it, maybe in denial that this disease is the real cause of her "forgetfulness". After all, if this were the case, there

---

[18] Ibid

[19] Ibid.

are terrible times ahead for Clara and her family members and friends, assuming of course they are the kind that love Clara and will be ready to do what is necessary to help her.

Many family members do not discuss the possibility of Alzheimer's disease with their loved one at all, as if avoiding the subject will make it go away. Likewise, the person experiencing memory loss is likely fearful of sharing this information with anyone, predicting that the future holds only loss —of identity, home, personality, memories, and eventually life. This dangerous mixture of denial by both sides lends to delay of diagnosis, preparation for the future, and treatment.

SIGNS OF MENTAL DECLINE: POSSIBLE INDICATORS OF ALZHEIMER'S DISEASE

Using Technology

As time goes on, family members become increasingly concerned that it may be more than old age affecting Clara's forgetfulness. It is often difficult to know the difference. The family visits Clara from time to time and begin to notice things that are out of Clara's normal way of living. For example, on repeated visits, they find her watching the same TV channel. When asked, why she's not watching one of her favorite programs on another channel, she replies that she likes this channel. Is Clara telling the truth or is she too embarrassed to tell them that she no longer knows how to use the remote - or is she even able to grasp the concept of multiple channels?

Other devices such a telephones—landline or cellular-- might now go unused. If she has historically used a computer to email or chat, she may no longer recall how to use these forms of communication. When her family member or friends ask her, she may say that she is busy or doesn't like to use such technology any more. Is she telling the truth or does she no longer know how to operate such technology devices?

> *Jan is an 82-year-old widow who lives in an apartment attached to her daughter's home. Her daughter recently had a landline installed in Jan's apartment, since Jan could no longer remember how to use her cellphone. Since installing the landline, Jan had neglected to answer even one call, claiming that "it must not be plugged in" or "I've never heard that phone ring; you must have brought me a broken phone".*

Maybe she still knows how to use a computer and get online. She probably receives, like most of us, advertisements. Does she reply to them, ordering items that are obviously not needed that are found lying around the house in unopened boxes, or opened but not used? Does she get defensive when asked about it?

Cooking

When asked what she had for breakfast, she says cereal. That alone is not an indicator but what if in the past she always talked about eating eggs and bacon everyday for breakfast. Maybe she doesn't know how to fry eggs or bacon any more. For dinner, maybe she is always eating boiled corn-on-the-cob. She may say she's no longer that hungry or she likes it. Truth may be that she no longer knows how to cook or even how to turn on the stove. Many homes for the elderly and even "over 55" communities have electric ranges as they in general are safer than gas ranges.

Mail

Are mail, newspapers, or advertisements piled outside Clara's front door or in her mailbox indicating that she forgot to check for mail on a daily basis? If Clara has a locked mail box near her home, it is a good idea to check it for her. If her relative asked her if she checked her mail, she may think she has and says so. She may say there's nothing but junk mail and does not directly answer the question.

Maybe a better way is to not confront Clara at all because she may become defensive. Maybe, say "Clara, I'm going for a short walk or I need to check my car, let me have your mail box key and I'll get your mail so you won't have to go get it". If inclement weather, maybe say, "Clara, the weather is bad. I'll get your mail for you as don't want you to catch a cold."

If a relative sees Clara's mail stacked on a table unopened or in a trash can unopened, is it because it's just "junk mail"? By looking at the envelopes, one can sometimes determine if some mail is important, like an overdue bill.

Finances

Some financial institutions may not know that Clara is in the beginning stages of Alzheimer's (or care, for that matter) and may talk her into a new credit card, opening a safe deposit box, opening one or more checking accounts. Surely they would not try to take advantage of Clara at her age, or would they? Some financial institutions may even send her an unsolicited credit card and charge her an annual fee for it.

> *When a loved one took over the finances of his mother, he found that she had three checking accounts opened, two with the minimum amount of money to maintain those accounts open. She also had a savings account containing only a few dollars, two CD's, several credit cards, debit cards, and a safe deposit box that was empty.*

If the mail or other documents are lying around, it is a good idea to at least look at the envelopes for signs of overdue bills, as a minimum. Some checking account notices may show returned for insufficient funds. In Clara's case, she has been writing checks using checks from various accounts not remembering which is her primary account and which others may just have the minimum required amount of funds to open the account.

She once was a very frugal lady, but now she seems to be sending money to various groups and charities when she never did that before. That in itself of course is not a bad thing to do; however, it may be an indication of the onset of Alzheimer's.

You may ask her why she is sending checks to charities. She may deny having done this or get defensive and tell you it is none of your business. If Clara has Alzheimer's, her most recent memories are gone. She may have forgotten mailing the money.

Driving

Obviously, the older we get, the slower our reflexes. Driving safety is always a great concern for all of us. It is more so of course when evaluating your loved one's potential mental issues, especially as they become older. If you, as Clara's loved one or friend, see potential signs of mental deterioration, it may be time to stop her from driving.

Does she tend to get lost driving, can't find her car in the parking lot, or runs out of gas? Of course, none of that is in itself unusual. However, maybe she ran or runs out of gas because she doesn't know how to work the gas pumps or pay for gas.

As Clara's loved one, should you take away Clara's car as it is too dangerous for her and other drivers for her to be driving? This is very difficult of course as Clara is a proud and independent woman.

If she has an accident or gets lost while driving, this is an important sign that maybe she should not be driving. Clara will likely realize that this major life event – the loss of driving ability – indicates her nearing the end of her life. Talking to her about losing her driving ability must be presented in a way that still allows her to keep her pride and ability to get around. After all, you are threatening her beloved independence.

Some communities and also retirement communities have free bus service and some have a driver who can take the person to the doctor or run other errands. Regardless of the different modes of transportation, this will still be a shock to someone like Clara; she very well may be in denial because she doesn't realize what is happening to her as Alzheimer's takes hold of her mental functioning.

So, when it is time to take away that car, it must be presented in as positive a way as possible. No matter what, it will not be well-received in the majority of cases. As Clara's close relative would you be prepared to take her when she wants to go somewhere, be her "chauffeur" until she dies, or she no longer thinks of driving, or at least quits fighting the loss of her car and driving abilities? Are you prepared to have serious disruptions in your life to drive her where she wants to go?

When you, as her loved one, start putting all such incidents together, it could be an indication of the onset of Alzheimer's or another form of dementia.

DIFFERENCES BETWEEN OLD AGE AND ALZHEIMER'S

The are some very fine lines between old age and Alzheimer's and some not so fine lines. Look at the lists below and keep them in mind when visiting your loved ones:[20]

Typical age-related changes:

- Making a bad decision once in a while
  Example: choosing to drive to town for groceries when it is snowing outside and the news has advised to limit driving in the poor weather.

- Missing a monthly payment
  Example: not realizing that the envelope that came in the mail was a bill, and simply tossing it in the trash.

- Forgetting which day it is and remembering later
  Example: An excuse for missing a doctor's appointment. This happens, particularly in the older, retired people who may have limited social lives; they may claim that they have no reason to remember the day of the week.

- Sometimes forgetting which word to use

---

[20] Quoted from alz.org

Example: the person cannot find the right word for the television remote control, so refers to it (out of frustration) as the "channel changer".

- Losing things from time to time
  Example: the person cannot find their house key and searched their entire purse, only to find it in the pocket of their sweater.

NOTE: There is often a fine line between a "normal" person and one with Alzheimer's. However, the collation of such things tends towards potential for Alzheimer's.

Alzheimer's signs:

- Poor judgment and decision making
  Example: the electricity goes out during a storm, and the person cannot figure out what happened or what to do about it.

- Inability to manage a budget
  Example: the person may lose the ability to track when bills are due—this may range from forgetting to pay them, to not remembering to write a check, to not realizing that the written checks need to go in an envelope, with a stamp and get mailed.

- Losing track of the date or the season
  Example: a loved one gets up and dressed in a nice outfit and calls you, wondering where you are; you're late. When you ask them what they mean, they tell you that they're going to be late for Christmas Eve service at the church. It's January.

- Difficulty having a conversation
  Example: the person may have difficulty following a conversation and to cover up this fact, they may change the subject to something they recall. Also, they may "zone out" during a conversation and not really understand anything you've said.

- Misplacing things and being unable to retrace steps to locate them
  Example: the person may have lost their wallet. When asked about when the last time was that they saw it, they may adamantly tell you that it was in their pocket that very morning. They cannot recall putting it in their pocket, nor can they tell you where or when they would have removed it.

Alzheimer's Association says you should know the ten warning signs. These include:

1. Memory loss that disrupts daily life
2. Challenges in planning or solving problems
3. Difficulty completing familiar tasks
4. Confusion about time and place
5. Trouble understanding visual images and spatial relationships
6. New problems with words in speaking or writing
7. Misplacing things and losing the ability to retrace steps
8. Decreased or poor judgment
9. Withdrawal from work or social activities
10. Changes in mood or personality

Granted this list does not necessarily mean a person is getting or will get Alzheimer's but the more of these indicators/traits a person has, the more likely their chances of entering a stage of Alzheimer's.

> *If you or someone you care about is experiencing any of the ten warning signs, please take them to see a doctor to determine the cause. Some of the symptoms may be associated with an infection, exacerbation of a chronic disease, a hormonal imbalance, poor nutrition—any of these factors can be addressed and the symptoms improved. If these symptoms are not associated with a physical cause, then early diagnosis gives you a chance to seek treatment and plan for your loved one's future and death.*[21]

---

[21] Ibid

SUMMARY

Currently, there is no cure for Alzheimer's disease. Early detection makes a difference. Many of those afflicted and their friends and relations may be in denial. It is important to know the signs of potential Alzheimer's onset and take action accordingly.

Just when you thought you could get on with your own life after the kids were grown, you now have another "kid" to take care of. There's a saying (source unknown), "There comes a time when your parents become your children".

No matter what, dealing with an Alzheimer's victim is never easy — and will continue to get worse as the disease takes hold of your loved one. It is so important not to ignore the warning signs if you truly love that person.

Educate yourself on this disease, monitor the health of your loved one, and ensure that they see a doctor regularly (it is best if you also attend as they may not remember or may dismiss the doctor's advice. If they reject that idea, see the doctor immediately after your loved one's appointment to discuss what was said). Also remember that Alzheimer's does not only strike the elderly.

Reading this book is one of the first good steps to dealing with this disease.

CHAPTER 2-2

DIAGNOSIS

*...those memory complaints did not guarantee
a person would develop dementia...*
— R. Kryscio, UK's Sanders-Brown Center on Aging

INTRODUCTION

Based on the indicators previously noted, and observation of your loved one, you have become aware of the possibility that they may have some sort of dementia and it may be Alzheimer's disease. What's the next step?

Options

There are three basic options you, as her loved one, can take:

- Do nothing, hope for the best.
- Do some research, such as on the internet, and look for some self-help tests that you can use to help determine if there is a possibility that your loved one may have Alzheimer's. You must use caution as assuming you're not familiar with the symptoms of this disease and therefore, mis-diagnosis is of course possible. While at the same time, it may help you in dealing with your loved one in the event you decide to take them to a medical doctor for a formal

evaluation. One approach is to make it into a game for both of you or all the family members.
- Take your loved one to a medical professional for evaluation without "self-testing".

> *The Self-Administered Gerocognitive Exam (SAGE) is designed to detect early signs of cognitive, memory or thinking impairments. It evaluates your thinking abilities and helps physicians to know how well your brain is working.*[22]

MMSE

"Folstein Mini-Mental Status Examination (MMSE)" is one test that has been used to evaluate a person with Alzheimer's symptoms and can be given over time, then comparing the differences in the response.

Some of the MMSE includes:

- Score a point for every correct answer when asking your loved one the year, season, date, weekday, and month.
- Another would be the name of the state, county, city, address and floor.
- Say three words and see if your loved one repeats them correctly.
- Ask them to subtract from 100 by 7.
- Ask them to spell a word backwards.
- Show them some items such as a pencil, watch, etc. and ask them to name the objects.
- Have loved one repeat the following phase correctly: "No if's and or buts".

---

[22] http://medicalcenter.osu.edu/patientcare/healthcare_services/alzheimers/sage-test/Pages/Index.aspx

- Ask them to write a complete sentence.
- Show the loved one some simple diagrams and then ask them to copy the diagrams on a separate sheet of paper.

These are just some of the examination questions that can be asked.

There are several examinations/tests that can be self-administered. Two examples are the St. Louis University Mental Status Examination, or SLUMS and the Montreal Cognitive Assessment, or MoCA. These exams should not take the place of a formal examination by a doctor; however, they may help in providing you with some information as to what your next steps should be.

If you determine by your own evaluation that your loved one experiencing nothing more than old age and not Alzheimer's or any other form of dementia, it may put your mind at ease; however, can you really be sure? After all, you may note that there are some similarities between the two.

> *Our recommendation is to get your loved one to a medical professional for an evaluation and maybe part of a general health examination.*

The Next Step

The next step then is to determine how to get that person to a doctor for a formal evaluation. In making that plan, you must consider that person may say there is nothing wrong and will argue against such an examination. So, it is best not to start by offering a potentially confrontational approach. Of course, you know your loved one better than anyone, so you know that person's mind better than others and therefore, know the best approach to take.

In general, the best approach may be to convince your loved one to get a physical since they are getting older and best to be in a preventive mode. This is especially true of one who has pride in independence, being healthy. You should not mention anything about any mental evaluation to determine if Alzheimer's mental decline has begun.

Remember it is also important that you accompany your loved one into the doctor's office. This is important so that you can hear what the doctor says and not wait outside waiting for your loved one to come out and explain to you what occurred. Chances are your loved one won't remember in sufficient detail or if Alzheimer's mental decline has begun, they may not even remember what the doctor said. In that case, the loved one may say, I'm fine or I don't remember but I'm ok. Either way, your loved one may be hiding the fact that the doctor discussed Alzheimer's and your they don't want to discuss it as a defensive mechanism. Your loved one may truly not recall what was said.

> *Being in the room when the doctor exams your loved one is vitally important. If your loved one is emphatic that you are not in the room, make arrangements to talk to the doctor after the examination whether your loved one is in the room or not.*

In an article[23] entitled: "Many Older Adults Don't Seek Help For This Condition…"[24], the holiday season such as Thanksgiving, provides an opportunity for family members to observe and discuss the mental and physical health of a loved one. Donald Rebhun, M.D., medical director of HealthCare Partners, believes that diagnosis is vital as early identification of Alzheimer's would lead to early treatment.

> *In a study noted in the article, 55% who had a form of dementia did not talk to a doctor; that means almost 1.8 million people had not been professionally evaluated.*
>
> *About one in seven of U.S. adults age 71 and older have some form of dementia and usually affects people age 60 and older, with risk increasing with age[25]*

---

[23] See https://www.yahoo.com/health/many-older-adults-don't-seek-help-for-this-103558365907.html
[24] By Amy Rushlow, Contributing Writer, November 26, 2014
[25] Ibid.

R. Kryscio talked about what a doctor can do for your loved one. For one thing, he said, a doctor can look at the possibility that other issues may be causing a loss of memory, such as vitamin deficiencies such as B-12.[26]

If you've done your research on Alzheimer's symptoms and documented you observations of your loved one over time, it is important to share such information with their doctor. This helps the doctor better understand their potential issues. This is most important if your loved one has not visited the doctor for some time, if ever. If so, there is no history from which the doctor can draw to compare the current state of the loved one's mind with past evaluations.

> *Depression, sleeping disorders, thyroid disorders, alcohol abuse, anxiety, and medication side effects can all cause symptoms similar to those of dementia, according to D. Rebhun, M.D., Medical Director, HealthCare Partners.*[27]

SUMMARY

You can take several steps to determine if your loved one should be immediately examined by a doctor, or after you conduct some examinations—tests first.

Getting a diagnosis may seem like a long process, that requires the close attention of family members to note the subtle (and sometimes not-so-subtle) changes in a loved one's memory and behavior. This process is an important one and will win in determining the next step.

You can start with a simple memory test, which you can easily offer at home and, at the very least, use the results as a starting point where you can re-test and see if your loved one's mental status has or declined in a few months' time.

---

[26] http://www.foxnews.com/health/2014/10/03/how-many=memory-slips-are-too-many/?intcmp=features/

[27] https://www.yahoo.com/health/many-older-adults-dont-seek-help-for-this-103558365907.html

DR. KOVACICH AND DR. ANENSEN-MCNEALLEY

Ideally, take the results of the memory test and visit the doctor with your loved one. Express your concerns clearly and with detail and examples, so the doctor can make the most accurate diagnosis and help you and your loved one maneuver through the potential treatment options, based on the exam's outcome.

CHAPTER 2-3

DIAGNOSED WITH ALZHEIMER'S — NOW WHAT DO YOU DO?

"If you learn to listen for clues as to how I feel instead of what I say, you will be able to understand me much better.[28]"
— Mara Botonis

INTRODUCTION

This chapter talks about what to do with your loved one once diagnosed with Alzheimer's.

Life Expectancy

As with most statistics, they are dependent on sources, history, and such. This applies to Alzheimer's disease as much as any matter where statistics are used. However, they are a good indicator for planning purposes when devising a roadmap for your loved one's battle with Alzheimer's. Not surprisingly, although the majority of Alzheimer's sufferers are over 65, in the United States, there may be more many who are under 65.

---

[28] alzheimers.net

> *Dr. Gerald L. Kovacich's mother was diagnosed with Alzheimer's in June 2011 and died in August 2014 - age 94 years, seven months. In looking back, there were signs of the disease years before the diagnosis. He failed to see them.*

It is important to always look for signs of Alzheimer's as we and our loved ones get older. We should not be complacent in thinking that it can't happen to us or to our loved ones until after they are 65 — if ever.

In preparation for caring for that loved one who has developed Alzheimer's, statistics indicate that they live an average of eight years. However, the lifespan after an Alzheimer's diagnosis can range from between four years and 20 years.

Be the Caregiver for Your Loved One, Find a Professional, or Find a Proper Facility?

Your loved one has just been diagnosed by a doctor and found to have Alzheimer's disease. Probably one of the most important and difficult decisions you will make in your life is "what now"?

If in the initial stage of Alzheimer's or close to it:

- Maybe the person can stay in their home with you or other trusted family members or friends doing daily checks
- Maybe you can have a caregiver visit daily
- Maybe you want to have a caregiver move into the home of your loved one
- Your loved one can move into your home or that of another family member

The decision will depend on several factors to include what you are willing to do and also what you can convince your loved one to do.

Are you prepared to have serious disruptions in your life to not only drive your loved one where they want—need to go, but also to help them with their finances, ensure they eat properly, bathe and such? Much depends not only on your willingness, but also your capability. It may be that you're willing; however, how do you prepare and are willing to handle some of the following as Alzheimer's takes hold over time?

Finances

Of course, one of the most important things to consider is finances. Considering the following:

- Assuming you decide it is best that your loved one moves into a facility. Does your loved one have the finances to live in an assisted living or other facility that cares for people living with Alzheimer's?
- If not, are you willing and capable of providing that financial support?
- If not, it is important to research government agencies that may assist in that financial support.

Availability

It is important, in coordination with the doctor, to continuously evaluate your loved one's status. When first diagnosed, as stated above: your loved one may be able to live alone with your periodic visits to ensure all is well and continuously look for signs of degrading mental and/or physical condition.

- Will you or someone you trust be available to be with your loved one 24/7?
- If left alone, can you trust that your loved one does not wander off, hurt themselves in a fall, accidentally start a fire, etc.?

Residence

Whether your loved one lives alone or with a family member is something to consider. This situation may change over time, either with the disease's progression or, oftentimes, with one harrowing event that prompts a move.

Regardless of the current living arrangements, consider the safety and security of the situation as it currently is, and anticipate what it might become as your loved one's disease progresses. Considerations include:

- Will your loved one live with you or you with them? Who decides?
- What if your loved one does not want to live with you or you with them?
- What will you do if your loved one does not want to live with you or for you to move into their home?

<u>Meals</u>

People living with Alzheimer's often experience changes in eating patterns for various reasons. The person may forget to eat, or forget that they've already eaten (and eat again). They may not know how to work the stove or oven anymore, or not realize that the refrigerator holds food and therefore don't look inside. They may, as the disease progresses, no longer realize what hunger pangs mean, and therefore they do not connect food with alleviating this feeling. They may forget how to eat or drink, and therefore set aside the use of forks and spoons and rather eat with their hands. Regardless of how the disease impacts your loved one, consider the following questions as they relate to meals and meal planning:

- Are you prepared to fix every meal for that person?
- What will you do if the loved one does not like what you fixed?
- Will you fix something different or say eat it or don't eat?
- Will you fix meals ensuring that the loved one gets all the vitamins and such that the doctor recommends?
- How will you ensure they eat?
- Sit near them as if the person was a child? - Because as the old saying goes: "Your parents become your children".[29]

---

[29] Source unknown.

> *Help is available. For example: If you or a loved one has been diagnosed with Alzheimer's or a related dementia, you are not alone. The Alzheimer's Association is one of the resources for information, education, referral and support to millions of people affected by the disease.*[30]

Medicines

Just like before your loved one had Alzheimer's, they have the right to choose how to live their life. They may choose to take medications or not. Likewise, they may have forgotten what the medications are, or what they are for, when they are supposed to be taken, and how many to take. Medications can certainly help your loved one, but if taken incorrectly they can cause great harm. It's important, then, to consider the following questions:

- How can you be sure that your loved ones takes the medicines in accordance with doctor's instructions?
- What will you do if you find that your loved one is not taking their medicines as required, even if they say that they are being properly taken?
- Will you give the drugs to them to take?
- What will you do if your loved one says that they don't want or need them?

You may find yourself filling prescriptions at the pharmacy, setting up an easy-to-use medication system for your loved one (such as the "morning, noon, dinner, bedtime" containers)., reminding them to take their medications at the right times, and following up to see that they indeed took them. You may find yourself at the doctor's office with your loved one, asking the tough questions like, "Does she need this medication?" Or "Is there a medication that can help with [symptom or behavior]?"

---

[30] alz.org

Entertainment

Boredom is certainly something to be concerned about with Alzheimer's, particularly since a lack of memory means your loved one may not recall hobbies enjoyed in the past, or even how to do them. People with Alzheimer's also tend to have short attention spans, meaning their involvement in anything entertaining may not last beyond five or ten minutes. Some questions to consider include:

- How will your loved one entertain themselves since over time their short-term memory declines and they can't read or even follow a story on the television?
- What will you do to try to stimulate your loved one's mind?
- Will you just leave the television on and sit them in front of it?
- What did your loved one used to like to do? This may include hobbies, chores, as well as work-related activities.

Transportation

It's likely that someone will have to eventually manage all aspects of your loved one's transportation needs, either due to the inability to drive or, more likely, the inability to plan a trip and get there on time. Some considerations include:

- Are you in a position to drive your loved one to various places such as the doctor's office, or to visit relatives and friends?
- How will you handle emergencies requiring immediate transportation?
- Do you have a list of emergency numbers, e.g. hospital, doctor, fire department, police, etc
- Will you use a transportation service and rely on the loved one to get where they need to go and get back without getting lost, forgetting their destination or home address?

THE ALZHEIMER'S PLAN

Making the Decision

This may be one of the most difficult decisions you may have to make in your lifetime. It will certainly be one of the most important decisions regarding your loved one.

After you've done your research and looked at all that is entailed in taking care of a loved one, you must eventually make a decision. Delaying it is a possibility depending on the mental state and also on physical status of your help in caregiving.

At first, it may be possible to just be more aware of their Alzheimer's condition and check on her periodically. However, at the same time, you should look at your loved one's deterioration and set certain milestones of decline. Based on those milestones, you will need to take certain action. For example, maybe have a relative or friend spend more time with the loved one. Further decline and she moves in with you or vice versa.

At some point, your milestone requires that a full-time or part-time Alzheimer's specialist be hired to spend time with your loved one. That may suffice; however, if it does not, then you should be prepared to move your loved one into some facility that has professionals who are experienced in dealing with Alzheimer's-related residents on a 24/7 basis.

If you are not willing or able to take on this tremendous task or hire a specialist to stay with them, then you will need to find a facility that can do what is best for your loved one depending on their condition.

SUMMARY

Alzheimer's disease impacts millions of people each year and it is not only the senior people who it will impact, although they form the majority of cases. It will of course affect all their friends and family members too.

There are several decisions to ensure your that loved one is properly cared for as this disease takes its toll and memory begins to decline over time. Proper care may begin with home care, to live-in caretaker, to a specialist facility.

No matter what you decide, it will take a great deal of your time and financial support may also be required.

Proper planning will save time, effort, and ensure that your loved one is treated with the love, compassion, and patience that they deserve.

SECTION 3

FINDING THE BEST FACILITY

This section is divided into five chapters:

Chapter 3-1: <u>Various Types of Facilities</u> defines the four main types of facilities and deciding on which type of facility where your loved one should reside.

Chapter 3-2: <u>Finding the Best Facility for Your Loved One</u> discusses how to go about finding the best facility for your loved one in the event that is what is needed and best for that loved one; financial independence versus financial assistance, types available; unique male and female needs.

Chapter 3-3: <u>Applications, Admissions, Screenings and Orientations</u> as the title states, this chapter discusses applying for your loved one's admission to the facility, general screening process, testing, independence versus independents, needing assistance; explains in general terms what can be expected once it is decided that the loved one is better off in an assisted-living facility or similar facility.

Chapter 3-4: <u>Facility's Staff</u> talks about what you should know about the staff at the facility where your loved one will reside; such as education, experience, personality traits, interviewing management and selected staff.

Chapter 3-5: <u>Other Residents</u> discusses the type of other residents you and your loved ones may encounter and how to interact with them.

CHAPTER 3-1

VARIOUS TYPES OF FACILITIES

"Always remember their life when they can no longer remember."[31]
— *Unknown*

INTRODUCTION

There are various types of facilities that care for people living with Alzheimer's. This chapter describes your possible options.

Caring with Options

There are so many people who are torn between exhaustion, caring for a loved one living with Alzheimer's or another form of dementia, and guilt for even considering a long-term care facility in which to place their loved one. The idea of placing the care, security and safety of a loved one in the hands of strangers may be terrifying. Careful consideration is imperative to an ultimate decision. The first thing to address is the type of facility available.

There are wide variations of long-term care options for people with Alzheimer's, and those variations are based on individual state regulations and emphasis (or lack of) by the state's legislature.

---

[31] Based on alzheimers.net quote

Informal Caregiving

Caring for people living with Alzheimer's occurs in the home by a family member who receives no payment for their work. This is referred to as informal caregiving and seems to be most often done by the adult daughters and granddaughters of the person with Alzheimer's. This can be exhausting work that fundamentally changes the relationship between parent (or grandparent) and family caregiver. Most families begin their journey through the long-term care spectrum here, with an effort to keep the person at home.

Hire In-House Caregiver

From in-home care many families take baby steps toward outside care. Likely the family hires an in-home caregiver for a few hours a day. This option allows the family caregivers to have a break, away from the home. It is highly recommend that the private caregiver be carefully vetted; proper credentialing, references and a background check are all good considerations when choosing a caregiver. Also, "interview" the caregiver formally and watch how they interact with your loved one. Educating the caregiver on your loved one's habits, likes and dislikes, and common daily activities helps prepare the caregiver for excellent performance and positive outcomes for your loved one.

Perhaps adult day care is the first transition. Adult day care allows a limited break in the day for the primary family caregiver, and provides socialization for the person with Alzheimer's. Many adult day centers provide medication management and health-related care as well as activities that focus on individuals with memory impairment. They are typically staffed with credentialed caregivers, a nurse, and perhaps a social worker or other types of professionals. Adult day care varies in cost and may be available as stand-alone centers or within a long-term care facility, such as assisted living facility.

THE ALZHEIMER'S PLAN

Adult Foster/Family Home

Transitioning from smaller to larger settings, the next is the adult foster/family home (or sometimes called residential-care facilities). The name varies by state, but these settings typically are family homes that are owned and operated by a person who lives in the home with the residents.

> *In Washington State, adult family homes can care for up to six individuals. The trained and certified care staff are what is typically deemed as universal workers – they provide hands-on care but also cook, clean, engage residents in activities, and provide prescribed medications. The adult foster/family homes are overseen by the state and must meet standard regulations for resident monitoring as well as environmental and staff training requirements. Most states that offer this type of long-term care require the owner and care staff to pass background checks.*

Assisted Living Facilities

Assisted-living facilities are larger and many have secure, dedicated units for the care of residents with Alzheimer's disease and other dementias. Many of these settings have robust activities programs that are designed to engage the person with Alzheimer's. Likewise, these secure units typically entice curiosity throughout the environment, like outdoor walking paths and "work stations" that invite residents to relive previous experiences.

Over the years assisted-living facilities have enhanced their services; many include transportation, onsite doctors or nurse practitioners, physical and occupational therapy options, and/or social services.

Many people with Alzheimer's live in an insecure assisted-living facility, meaning that the front door is unlocked and the residents are free to come and go. As long as the person with Alzheimer's does not have a tendency to wander, this type of setting may be perfectly fine.

Most assisted-living facilities have nurses to oversee the care and services residents receive. Qualified, trained caregivers provide hands-on care. These larger facilities generally feature large dining rooms with dedicated kitchen staff to prepare and serve meals. Each state has its own set of regulations to oversee assisted-living facilities, and many states limit the type of care as well as the type of resident that can live there.

Nursing Home

Perhaps the most well known type of long-term care setting is the nursing home, or skilled nursing facility. All 50 states have this option. These facilities are routinely inspected not only by the individual states but by the federal government as well. Nursing homes feature round-the-clock nursing care in a more institutional setting.

Typical nursing homes also feature rehabilitation services such as physical, occupational, and speech therapies. Many states, due to the exorbitant cost of care in these settings, have emphasized care for those with Alzheimer's and other dementias in less restrictive and less costly alternative settings, such as assisted livings and adult foster/family homes.

Palliative Care

Relatively new to the medical world, palliative care offers an additional layer of support to people living with a chronic disease like Alzheimer's. This type of care features doctors, nurses and other healthcare professionals, working as a team, to promote the patient's quality of life. Comfort becomes the primary focus in this type of care, and supports family in addition to the person living with Alzheimer's. This type of care is not exclusive; it works in conjunction with the person's primary care provided and can blend in harmoniously wherever the person is living.

Hospice Care

Hospice care is a benefit that all people on Medicare can access, provided specific health-related requirements are met. The person must be deemed terminal, with an expected death within the coming six months. Hospice care focuses primarily on the comfort of the dying person, with close inclusion of the family.

Hospice care can be provided anywhere including the person's home or in a long-term care setting. Hospice includes licensed nurse visits, which typically occur one to two times a week and may increase as the person's health declines.

Spiritual counseling is available should the dying person and/or the family request it. A bath aide also is available to provide routine bathing services for the individual. Hospice care also includes the provision of a hospital bed and special mattress to promote comfort, oxygen should the person need it, and other person-specific items that can promote comfort during the dying process. Certain medications are prescribed and given to manage pain as well as shortness of breath and other symptoms that accompany the dying process.

SUMMARY

The type of facility to choose is just a starting point; there are many other factors to consider as you move through this process. You may find yourself researching more about what is available in the state where your loved one lives – each state has a government agency where you can learn more about long-term care options; there are also trade associations representing long-term care facilities where you can also find out more before making your decision.

CHAPTER 3-2

FINDING THE BEST FACILITY FOR YOUR LOVED ONE

Often hear people say that a person suffering from Alzheimer's is not the person they knew. I wonder to myself - Who are they then?[32]
—Bob DeMarco

INTRODUCTION

This chapter addresses the process of finding the best facility that meets the needs of your loved one.

Shopping for the Care Facility

Let's face it – no one wants to "shop" for a long-term care facility. It has been ingrained in society's brain that the nursing homes of yesteryear are alive and well today, and that going to one is akin to a death sentence. This could not be further from the truth. While you may have never thought in a million years that you'd be looking for a care facility for your loved one, the reality is – caregiving decisions are difficult but necessary.

---

[32] alzheimersreadingroom.com

Crossing the line from family member to caregiver is a tough path. What used to be private, personal activities like toileting and bathing all of the sudden become a group effort.

> *Dr. Anensen-McNealley was the primary caregiver for her grandfather after he had hip surgery; he also had dementia. Showering him was not only uncomfortable for her, but mortifying for him. Add to that the toileting (and cleaning up his private parts) and she found her depressed and overwhelmed. She had lost her role as beloved granddaughter and she became the hated nurse. It was time to reach out to an assisted living community for help. It was the best decision my family could have made.*

There are so many factors that play into this huge decision. We urge you to consider them now, BEFORE your loved one needs a care setting. Waiting only makes a decision rushed and it may end up in a situation where your loved one has to move again.

When it comes to people living with Alzheimer's, the term "transfer trauma" is real – where changes in one's environment prompt challenging behaviors, depression, and overall health decline. *So make the right decision the first time.*

There are geographical and physical issues to consider:

- How far away is the facility from family? As a busy working adult, you may think, "it doesn't matter how far the facility is from my house. I'll make time to visit." Reality check – the further away you are from the facility, the less often you'll probably visit.
- Does your loved one enjoy solitude or crowds? This is a big deal. There are assisted- living facilities that are more like huge hotels than apartment-like homes; residents eat in large dining rooms with upwards of 50 or more people. Multi-level buildings may be more like a maze than a home too – how far would your loved one need to walk to get to dinner or an activity/event? On the flip side, there are small, intimate adult family/foster homes with only a handful of seniors living there; if your loved one prefers more of

a home-like setting with family dining and shared living space, this too is an option.

- <u>Private apartment or shared room?</u> Many nursing homes have shared institutional-like rooms with a shared bathroom and nothing more than a bed, closet, and chair. Most assisted livings feature private apartments, but not always – research has shown that, for people living with advanced Alzheimer's, living alone is not always a good thing. For this reason, many dedicated memory care units feature shared rooms and maybe even shared bathrooms and showers. What is important to consider in this situation is the fact that your loved one likely will not spend much time in the bedroom —a great facility that caters to people living with Alzheimer's disease offers consistent activities and events that pull the person from a secluded life to a more social setting, where they can be observed and interact with others. Keeping the mind and body active limits the person's time in their room, whether shared or private, to personal care, bathing and sleeping.
- <u>Is there ample common space, both inside and outside?</u> Areas to visit and socialize, as well as engage in hobbies such as gardening, crafting, and exercise are all important to a well-rounded living situation.
- <u>Are the doors locked?</u> Depending on your loved one's needs and level of dementia, it may be best to find a facility that features a secure egress system. This way, your loved one will receive care in a secure environment where fear of him/her leaving the building unattended can be minimized. This feature may not be warranted, however, if your loved one does not tend to wander. Always side on security-safety.
- <u>Costs a factor?</u> Ideally, you would like your loved one in the best possible facility; however, costs usually increase as the care provided increases. How much can you and your loved one's estate afford and for how long? Some long-term care facilities serve people on the Medicaid program. If your loved one does not have a lot of private funds you need to ask facilities what their policy is on accepting Medicaid as a payment source.

> *In Washington State, an assisted living facility or example, may provide individual rooms in a safe and secure setting with medicines and meals being provided under 24/7 staff supervision. The costs may be about $5000 per month with less than 40 residents in the facility. Those facilities also provide cleaning, activities in and out of the facility.*

Before selecting a care setting for your loved one, it is important to visit it first. We encourage you to drop in several times – day shift during the week when all of the management team is there, on a weekend, evening shift during the week and weekend, and perhaps if you have the opportunity, visit on a holiday.

Find an activity on the facility's calendar that your loved one might enjoy, and plan on visiting then so that they can participate. Eat a meal there (or two or three!) to ensure that your loved one likes the food. If special diet is required, that must be discussed with the management. There may be extra costs associated with that.

There are some important questions you should have ready when talking with the director and staff. These questions can make or break your decision. Remember, you are looking out for the safety and quality of care and life of your loved one!

- <u>Can I see your last state survey?</u> As a past state surveyor for assisted living facilities in the state of Washington (Dr. Anensen-McNealley) I can assure you that reading these documents can be scary. The requirements for documenting compliance issues outlines the worst-case scenario. Read, ask questions for clarification if necessary. Ask how they fixed any cited issues, and how they plan to prevent them from recurring.

> *In Washington State the inspection report (called a "Statement of Deficiencies") uses terms like "failed practice" and "placed residents at risk of harm" for varying issues that may or may not place a resident at risk of harm (e.g., storing a mop in a non-vented storage room may warrant a written report that states this action places residents at risk of harm. The same statement may be made in the document when the facility does not take appropriate action when a resident with congestive heart failure has difficulty breathing).*

- <u>What is the policy on Medicaid?</u> If your loved one will be paying privately but is on limited funds, it may be worthwhile to select a care setting that will allow them to transition onto Medicaid if money runs out. That way, you won't need to shop for another place later.
- <u>How often is a nurse here?</u> Nursing homes are required by federal law to have licensed nurses 24 hours a day, seven days a week. As for the other care settings, it all depends on state regulation, the facility/company's philosophy of care, and financial limitations. In many assisted-living settings, a nurse is there 40 hours a week. The role of the nurse is also something worth noting – the nurse may serve solely as a manager, supervising care staff and following up on residents' illnesses and doctors' appointments. Some smaller adult family/foster homes have nurses who own the company, and may live in the home and be available all the time; others still may have a nurse on call but rarely in the home.
- <u>What are some reasons my loved one may need to move out?</u> This is also something to consider. All long-term care facilities have limits to their caregiving abilities. These limitations may be state-imposed or facility/company-imposed. Some state regulations, for example, don't allow certain facilities to care for people who take insulin and cannot self-inject it. Some companies/facilities limit their care to a certain population of residents or level of care; for example a company may say "no" to a resident who needs two people to aid in getting them out of bed or moving from a chair

to the bed. As your loved one declines in health and cognitive functioning, it is important for you to know at what point a higher level of care may be needed.

- <u>Who owns the company?</u> You may be interested in knowing if the company is a large chain or a small 'mom-and-pop.' There are pros and cons to each, of course. This is simply personal preference but may very well influence your decision. Likewise, regional versus national companies have varying levels of support and internal oversight.
- <u>What are the expectations of the family?</u> The facility may not have transportation services or anyone to take your loved one to a doctor's appointment. They may not offer three meals a day, and expect family to ensure their loved one has groceries. They may not get medications from the pharmacy for your loved one (or on the opposite end of the spectrum, they may not allow families to assist with medications at all). Find out what your role will be once your loved one moves in, and plan for that.
- <u>Are religious factors important?</u> Some facilities may have a religious orientation. They may take all eligible applicants; however, they may have Bible classes, church services, etc. If your loved one is not of the same faith or is not religious, you may want to determine if they encourage attendance, move your loved ones to that area without asking permission. Best to find out these things in advance and decide what to do if there are conflicts of interest.

We are aware of one facility that encouraged a Christian environment, held Bible classes and services on Sunday. A resident was not religious and declined to attend. Generally, after a resident dies, the other residents gather and talk about the death of that resident in a loving and often humorous way. The resident who died and her family were not allowed to have such a meeting unless it was of a religious nature with prayers.
The family declined.

As you consider current or future care needs of your loved one, it is important to explore long term care options in your state and how that care can be paid for, given your loved one's financial resources.

Long-term care options are unique to each state, with the federal government overseeing nursing homes and, to a lesser extent, home- and community-based waivers for less restrictive (and less expensive) care. Funding of long-term care is also limited and primarily focuses on private resources.

- <u>Medicare Part A</u> can be used for a limited stay in a nursing home following a hospitalization. Medicare also funds hospice care and therapy (physical, occupational, and speech) regardless of where the person lives.
- <u>Medicaid</u> can be used to pay for many long-term care options including nursing home care. To a lesser extent, Medicaid is available to pay for some or all of the care in assisted living, adult foster/family homes, adult day care/health centers, and in-home care.
- <u>Veteran's Administration</u> pays for long term care; if your loved one served in a branch of the military, (VA) they may benefit from long-term care services that are partially or fully covered by the VA.
- <u>Long-term care insurance</u> pays for specific aspects of long-term care and depends on the policy.

Since each state has the option to dedicate public funds to the care of its elderly and disabled citizens, the options vary from state—to—state. Over the past ten years most states have branched out to offer more long-term care options for those citizens living with Alzheimer's.

Regardless of all the questions, the visits, the reviews and possible suggestions and recommendations from acquaintances, the ultimate decision lies with you and your loved one.

Internet search engines should be used to find reviews that could provide excellent information or a diatribe of complaints from a disgruntled or terribly unhappy person. Do your best to see through the shades and ultimately come to a decision about the best care setting for your family's situation.

> *In the case of Dr Anensen-McNealley's grandfather, our family (along with Grandpa) decided upon a 45-bed family-owned assisted-living facility just two miles from my house. It was not a five-star location by any means – the building had been there since the 1980's and while it wasn't opulent, my grandfather fit right in (he was a retired millworker). The care staff were loving. The facility overall did not perform outstandingly with their state surveys, but the care of my grandfather was top-notch. When he got to the point where he needed more care and supervision due to his advancing Lewy Body dementia, we tried out the secure memory care unit. It didn't suit him – sharing a room was not for him, and so we moved him back to the assisted living side of the building and filled in his extra care needs with family and additional caregiving support. "Going with the flow" and choosing what was most important both for my grandfather and our family, worked well.*

SUMMARY

Finding the right facility requires that you do your homework. Look at all the possible facilities and compare the pros and cons of each. Visit the various ones, talk to the director and staff. Make periodic visits at various times of day, days of the week. Talk to the residents and get their different opinions — keeping mind their state of mind. Review their inspection reports.

If you are barred from no-notice visits, discussions with staff and residents, reading the inspection reports, ask why? There may be a good reason or an indication to look elsewhere.

CHAPTER 3-3

APPLICATIONS, ADMISSIONS, SCREENINGS AND ORIENTATIONS

Caring for an Alzheimer's patient is a situation that can utterly consume the lives and well-being of the people giving care, just as the disorder consumes its victims.
— Leeza Gibbons[33]

INTRODUCTION

The decision has been made to move your loved one to a care setting. Depending on the particular facility and the owner's level of sophistication, the process will vary in length, paperwork, and processes. There are basics, however, that will apply to all care settings regardless of these factors.

Moving the Loved One

Moving a loved one into a care facility may seem akin to buying a used car. There are tons of papers to sign that are often overlooked and perhaps not even read by the purchaser. This is a terrible mistake – be sure to take the time to read *every word* and ask questions along the way. If your gut is telling

---

[33] https://www.brainyquote.com/topics/alzheimer

you something isn't right, then it probably isn't. Remember you are talking about the care of your loved one by strangers for probably the rest of their life.

> *Placing your loved one in a facility is a stressful experience and probably one of the biggest decisions you will make in your entire life. Treat it as such.*

An assessment or similar evaluation will take place in order to determine whether the potential resident qualifies to live at the community. Each long-term care facility has its limitations regarding the level of care they can provide, whether through regulation or philosophy (or both); the assessment will determine whether the potential resident falls within the realm of these expectations. Likewise, this assessment allows a time for each party to get to know one another. For example, the person completing the assessment (this may or may not be a nurse) might determine that a potential resident's behaviors do not "mesh" with the other residents currently living in the home; this may lead to denying move-in.

Some topics addressed during an assessment might include:

- Medical and mental health diagnoses, past surgeries and illnesses, etc.
- Level of assistance with dressing, grooming, toileting, bathing, etc.
- Medication review, with level of assistance with medication management
- Food and dining preferences, special diets
- Abilities and limitations when communicating
- Sleep habits
- Condition of the skin, particularly focused on any wounds or other chronic skin issues
- Nursing needs, such as tube feeding, injections, wound care, blood sugar monitoring
- Behaviors that place the person or others at risk
- Smoking

THE ALZHEIMER'S PLAN

- Activities/hobbies enjoyed
- Level of assistance with walking and transferring from one location to another

Once the assessment is completed and the person is identified as appropriate for move-in, the paperwork can commence. Most facilities require a deposit of some sort, particularly if your loved one will pay privately for care. If you're looking at a continuing care retirement community (a campus-like setting where independent, assisted, and skilled nursing living options are all on one property), you will be likely to be expected to pay a nonrefundable "entrance fee" that can be upwards of six figures, along with monthly fees.

Should you choose a traditional long term care option such as assisted living or adult foster/family home, the deposits are very similar to moving into an apartment or renting a house. If your loved one has a pet, there will be a deposit for that too. NOTE: Some may not allow pets.

If no pets of any kind are allowed, then what do you do with that pet? If they allow pets, is your loved one capable of caring for that pet if they may not be able to properly care for themselves? These will be factors for consideration; some long-term care facilities charge extra to care for your loved one's pet should they be unable to do so.

The facility management will likely explore your loved one's ability to pay; expect questions regarding income, retirement, insurance, and assets to ensure future payments will continue. You may be required to provide proof of the person's income prior to move-in.

Admission paperwork goes well beyond deposits and proof of income. Some items you will likely see and need to review include but are not limited to:

- <u>Admission agreement</u> – this typically includes specifics as to which apartment/room the person will occupy, expectations of the renter with regards to that apartment, and cost of care.
- <u>Residents rights</u> – a review of the rights residents enjoy in the facility should be provided. Typically the same rights you enjoy in your own home will be applicable in the facility.
- <u>Disclosure of services</u> – this might be folded into the admission agreement, and covers what services are available (and not

available). Likewise, this document should identify limitations of care that, should your loved one reach any of these limits, would warrant moving from the facility. Not all states require a document such as this, so please be sure to ask what the facility's limitations are.

- <u>Pharmacy agreement</u> – if the facility uses a "house pharmacy" you may be required to sign an agreement for the pharmacy to access your loved one's private health-related documents.
- <u>Photo release</u> – your loved one may be in a facility newsletter or on TV. A photo release is necessary to allow this.
- <u>Resident handbook</u> – most facilities have a resident handbook; this manual serves as an orientation guide to the new resident and spells out important information such as meal times, storing valuables, visiting hours, behavioral expectations, and more.
- <u>Arbitration agreements</u> – if you live in a state that allows facilities to enter into arbitration agreements with residents, this form may be presented to you. Many states have done away with them and the federal government no longer allows them in nursing homes/skilled nursing facilities. Know that you do not have to sign this form and simply remove it from the admission packet if you feel it warranted to do so.
- <u>Policies and procedures</u> – there might be upwards of 15 to 20 policies and procedures to review before your loved on moves in. If these are not presented to you, be sure to ask for them.
    o <u>CPR</u>. This policy outlines when and if staff honor a person's wishes regarding CPR.
    o <u>Smoking</u>. This policy outlines whether or not smoking is allowed and if so, what the limitations are.
    o <u>Cannabis</u>. If you live in a state where cannabis is legal, this policy should be presented to you. It outlines limitations of its use amongst residents in the facility.
    o <u>Assisted suicide</u>. If you live in a state where assisted suicide is legal, this policy should be presented to you. It outlines

THE ALZHEIMER'S PLAN

limitations of staff involvement as well as resident's abilities to implement assisted suicide.

- Pets. This policy should outline what type and size of pets a resident can have, if the facility allows them. It should also cover necessary veterinarian care, vaccinations, temperament, toileting, and additional costs, if any.
- Outside care. The facility will likely have expectations of healthcare workers you may contract with to provide care for your loved one. For example, if you hire a private caregiver to sit with your loved one at night, the facility may want that person to go through the facility's orientation, get a TB test, and be credentialed as a caregiver with the state.
- Grievances. This policy covers how residents may file grievances with management and management's expectation in responding to resident complaints.
- Medicaid policy. This policy covers when a person may transition to Medicaid and the facility's capacity for Medicaid. This is important, as your loved one may, in the future, need Medicaid to pay for some or all care, and if the facility to which you're moving your loved one does not accept Medicaid as a payment source, you'll likely have to move again.

Many long-term care facilities have formalized methods to orient new residents to their surroundings, while others are more informal. There are various ways that have proved successful in welcoming a new resident into the "family" of facility life. These might include:

- Welcome committee, where seasoned residents invite the new resident to attend a facility event and/or sit with them in the dining room for meals.
- "Buddy" system, where a seasoned resident is assigned to a new resident, and serves as a guide for the first few weeks.
- Escorts, where staff stops by a new resident's apartment and walks with them to and from meals, activities, etc. for the first few weeks.

Adjustment period

A move to a new location can be a harrowing experience, not only for your loved one living with Alzheimer's, but for you as well. Leaving your loved one in a long-term care facility can be compared to dropping your kid off at sleep-away camp for the first time. It wrenches at your heartstrings, especially when your loved one, confused and not understanding the process, begs you to stay or take them with you.

Professionals at the long-term facility understand that the first few weeks of a person's move there will be difficult. It takes time to find a new routine, to locate a new room and the dining room. The feeling of safety and security may be lost, and your loved one may demonstrate challenging behaviors because of the overwhelming change. Please know that this is normal, and your loved one will adjust, given time and support from you and the staff.

> *A 68-year-old retired auto-mechanic with Alzheimer's disease moved to a secure memory care facility unit after driving his truck from home and getting lost. His wife felt that she had no other option in keeping him safe and signed the paperwork and left shortly after getting him situated in his room. Within the hour, he began screaming at staff and other residents, throwing plates and cups, and attempting to climb the fence outside to "get the hell out of here." Staff immediately took action, talking with him about cars he had restored, and eventually he sat down with a cup of coffee and bragged for hours about one particular care that was his "pride and joy". The staff contacted his wife and got pictures of the car he had restored over the years, created a scrapbook. This book became his (and the staff's) "calming source" and the staff used these pictures to ensure he felt safe.*

SUMMARY

The move-in process should never be rushed, no matter the circumstances. Moving into a care setting is an important decision to make, and choosing the right one takes time and careful attention. Review the paperwork, ask questions (and get answers), and consider all options for making a final decision.

CHAPTER 3-4

FACILITY STAFF

The best way to find yourself is to lose yourself in the service of others[34].
— Mahatma Ghandi

INTRODUCTION

Excellent staff is the primary key to patient engagement and happiness. Whether your loved one is being cared for at home or in a community-based care setting, caregivers play a primary role in not only managing chronic illness, but minimizing the troubling behavior that often accompanies Alzheimer's disease and other dementias.

The Staff

The very basic trainings that seem to be required across all sectors include CPR and first aid. Most states require that caregivers have at least basic caregiving skills in order to work in assisted livings and adult foster/family homes. A credential as a nursing assistant is required for caregivers to work in a skilled nursing facility. This training is the bare minimum, for it takes a special person with dedicated training and insight to care for people who live with dementia.

Washington State appears to have the most robust and highly trained caregivers when compared to all other states. Caregivers working with

---

[34] Goodreads.com/quotes/

people in an assisted-living facility or adult foster/family home must be credentialed by the state as either a home care aide or a certified nursing assistant. In order for a caregiver to obtain one of these certifications, they must complete 75 hours (home care aide) or a minimum of 85 hours (nursing assistant) and pass state skills and written exams.

Caregivers in Washington who provide care to people with dementia must successfully complete a minimum of eight hours of training focused on dementia care and pass a state exam. While all caregivers must complete a minimum of 12 hours of continuing education each year, those who work in a dedicated memory care unit must ensure that six of those 12 hours are focused on dementia-related topics.

One of the very best pieces of education for any caregiver working with people who have dementia isn't a training at all – it is a 12-minute tour where the caregiver gets to experience what it feels like to have dementia. This program, developed by a company called Second Wind Dreams, has been dubbed "The Virtual Dementia Tour" - cannot say enough for its impact on a caregiver's professional growth.[35]

Training is just the beginning of the journey to becoming an excellent caregiver for patients living with dementia. Some of the best caregivers come from vastly different backgrounds; most did not start their careers in long term care but rather transitioned into it later in life. Personality, then, plays a huge part in caring for cognitively impaired individuals.

Perhaps the most important trait of an excellent caregiver is what is dubbed as "flair." The caregiver with "flair" has the ability to go with the flow and adjust their behavior and intentions to that of the resident's.

> *A caregiver needs to have a resident change a dirty shirt before a meal, and the confused resident does not realize that their shirt is soiled and thus resists the change by pushing the caregiver away and yelling at them. The caregiver with flair "accidentally" spills a drink on the resident's shirt and apologizes profusely. The resident, uncomfortable in a wet shirt, easily changes into a clean one with little to no fuss.*

---

[35] secondwind.org

THE ALZHEIMER'S PLAN

Some of the best caregivers we've trained and/or worked with were bartenders, waitresses, marketers, teachers, or other customer-based professionals in their past careers. Others still were fresh out of high school and simply had what it took to care for not only a resident's physical needs, but emotional, social, and spiritual needs as well. All of these caregivers shared traits worth mentioning here:

- The gift for gab: Anytime you are providing care for a resident, talking with them about regular life seems to create a friendship that promotes trust. The caregiver should also be able to start a conversation anytime – during a meal, during a game of cards, while gardening or writing a letter or doing just about anything with a resident – not just during care-related tasks.
- Flexibility: A to-do list may be a nice idea for any worker at the start of a workday. When caring for someone with dementia, however, that to-do list may go to the wayside entirely or at the very least not get done in the order or timeframe originally planned.
- Creativity. A person living with dementia is living in a different time, and perhaps even in a different place. A caregiver must be creative in not only getting things done, but also in getting the resident engaged in daily life. Coming up with ways to get the resident moving, eating, washing up, smiling and laughing are all the job of the caregiver, and it takes creativity!
- Thick skin: Residents with Alzheimer's may not be the nicest people all the time. The disease itself changes the brain, and oftentimes turns social filters to the "off" position. This prompts the resident to say whatever is on their mind, even if it's not socially acceptable, politically correct, or kind. Likewise, some symptoms of certain types of dementia change a person's personality inward, causing them to become self-centered and perhaps even egotistical. These residents will likely not say "thank you" or may be grumpy or aggressive with caregivers. A thick skin, where harmful words and phrases and physical outbursts are viewed as the disease process and not as the person's true self, serves to create an understanding caregiver.

- <u>Reading body language:</u> Many people living with dementia are unable to tell you what they are feeling or sensing. Instead, they may "act out" or behave in such a way as is difficult for the rest of the people around them. A great caregiver understands that behaviors and body language are just another way to communicate. The caregiver's job is to figure out what the person is trying to say, and take necessary action.
- <u>Super-sleuthing:</u> Many times there are issues with residents that are mysteries. Lost glasses or dentures that seem to have disappeared without a trace; a bruise that shows up out of nowhere, and the resident unable to tell you how it happened; a resident who hits out and yells at one particular caregiver, but is sweet as pie to all of the others. A great caregiver can put on their inquiry hat and investigate to get to the root of the matter. Solving mysteries should be in the job title!
- <u>Positive, fun attitude:</u> People living with Alzheimer's tend to "suck up" the moods of those around them. For that reason, excellent caregivers ensure that their shifts are upbeat, positive, and fun. Life is good, after all, and a resident in a good mood makes the day go by so much better.

While the environment is an important factor in selecting a care setting, the staff is the most important consideration that should prompt your decision. Be sure to not only speak with them, but observe them too.

Some questions to ask include:

- <u>What is the caregiver turnover rate?</u> High turnover negatively impacts resident care and confidence in staffs' performance. Be wary though — turnover rates are typically high in long-term care and so a high rate might not correlate to poor care. For many, a caregiving job might be a first job or a stepping stone to a career in healthcare. Many caregivers work and go to college

THE ALZHEIMER'S PLAN

or have aspirations to do more. Turnover, then, may be a natural movement that results in what appears to be constant change.
- <u>What kind of dementia/Alzheimer's training does the staff have?</u> This is a big factor, since the care staff will be assisting your loved one with the most intimate of activities. You'll want to be assured that the staff is fully trained and capable of managing difficult situations. Some companies ensure that their staff has the bare minimum that the state requires by way of training, while other companies have higher expectations and therefore additional educational opportunities.
- <u>What is your staffing model?</u> You should know how many care staff is working on each shift, and how often a nurse is there.

Some observations to make include:

- <u>Staff—resident interaction.</u> Is staff welcoming, positive, kind? Do they initiate conversation with residents? Do they call residents by name (as opposed to terms of endearment like "sweetie" and "honey")?
- <u>Staff—staff interaction.</u> Is staff talking to each other instead of talking with the residents? Do they talk about residents in front of other residents/visitors? This raises concern.
- <u>Engagement.</u> Is staff engaging residents in activities, chores, hobbies, etc or are they sitting off to the side of residents, looking at their phones or otherwise doing things that are not directly related to work?
- <u>Residents.</u> Do the residents appear happy? Clean? Well groomed? Are they sleeping or otherwise unengaged in their environment?

Interviewing Facility Staff Caregivers

If you have a chance to interview some staff at the facility where you are thinking about moving your loved one, you should consider evaluating them first:

- What is your professional background (not personal unless you want to provide it and it is relative) e.g. education?
- How did you prepare for this job?
- Why you chose this profession?
- In what related positions have you worked?
- Is it what you expected?
- What did you find that surprised you?
- How long have you worked in the profession?
- How many such places have you worked?
- How do you get prepared for such emotional tasks each day?
- How do you deal with their deaths?
- What happens when you get home?
- What do you do to unwind from the stress?
- Examples of what you have had to deal with?

Such questions and of course their answers will tell you a great deal about the caregivers who are going to be taking care of your loved one.

Remember though that their tasks are not easy ones. They may often be subject to stress and illnesses due to their job. For example:[36]

- Sleeping problems-sleeping too much or too little
- Change in eating habits-resulting in weight gain or loss
- Feeling tired or without energy most of the time
- Loss of interest in activities you used to enjoy such as going out with friends, walking or reading
- Easily irritated, angered, or saddened
- Frequent headaches, stomach aches, or other physical problems.

An interview with a caregiver captures information that allows for insight into the profession and frustrations that accompany that profession, for instance:

---

[36] http://us-mg4.mail.yahoo.com/neo/launch?.partner=ftr&.rand=51lc3vo8pnme

> *She got involved in the assisted-living profession by accident. She was 17 when she had to help her aunt take care of her uncle due to his Alzheimer's disease. She has been working in this profession now for about four years. She likes the profession and plans to get more education and focus more on children's care. Previously, she had attended junior college, taking general studies for about a year and a half, and then became a Certified Nursing Assistant (CNA).*
>
> *As part of her certifications for working in assisted-living facilities, she was required to also to be knowledgeable in food handling, English, physiology, chemistry, CPR, drug administration, taking vitals; however, such certifications do not allow her to actually put drugs in the resident's mouth. She can apply first aid such as a bandage on a cut but she is not allowed to take the bandage off. The LPN cannot delegate such authority to a CNA.*
>
> *She did not like some of the changes in the assisted living related laws. She believed that it was well-intentioned but had negative results such as not being able to attract some good candidates for assisted-living positions.*
>
> *The hardest part of the job is letting go when people die and of course there are often deaths as obviously the assisted living residents are all old. For most, it is their last living place on earth. The longer she had known the residents who died, the hardest it is to let go.*
>
> *Her way of relaxing after a strenuous day of work is driving the long way home, Yoga, meditation, and many cups of coffee.*

So, when you are interacting with caregivers, keep in mind what they are going through on a daily basis. Practicing compassion, patience and understanding when dealing with them would be appreciated by them.

SUMMARY

The above are just suggestions and a starting point. Care staff's hard skills – tasks that are imperative to the job and include measuring vital signs and assisting with activities of daily living – are important and necessary. But when caring for people living with Alzheimer's disease, the soft skills are most necessary: reading body language, anticipating the person's needs, listening, and communicating with compassion. Combining these two types of skills creates the perfect caregiver.

CHAPTER 3-5

OTHER RESIDENTS

...I miss my mother. Bits of her are still there. I miss her most when I'm sitting across from her.
— Candy Crowley[37]

INTRODUCTION

This chapter discusses the other residents that may be encountered in your loved one's facility. They will have an impact on your loved one as may talk, dine, visit, et al. together.

The Residents[38]

Depending on the type of facility you choose, there will be varying types of other residents also living there. Most often, particularly in assisted living, your loved one will have their own apartment. In other settings, though, there might be a roommate. While staff strive to connect similar folks to share rooms, this is not always possible. If your loved one does end up sharing a room with someone, it will be a person of the same sex.

One can compare assisted-living facilities to dormitories, in that a [sometimes large] group of strangers come together to share living and dining space for a cohesive reason. In the case of long-term care, that reason is to receive care and necessary services to promote one's independence.

---

[37] alzheimer.net
[38] Stories of some residents are described in Attachment 2

To take the dormitory analogy further, the likelihood that one's housemates are just like your loved one is rare indeed. This is what may be deemed as a "culture clash" where those of differing faiths, experiences, backgrounds, heritage, and habits come together in what hopefully is a homelike situation. Add to the mix a staff of equally differing backgrounds and what you may find is a cultural menagerie of the utmost proportions.

What the group will have in common include age, varying chronic illnesses, and the need for both emotional and physical support. Many will discover new friendships and interests that encourage living as opposed to simply existing.

As your loved one will likely be living in an environment with other people who also have Alzheimer's or another form of dementia, the interactions may become more complex (for you, your loved one, and for other residents). Because dementia robs a person of important memories, it may not be far-off to consider your role changing from daughter/son or granddaughter/grandson to that of a friend, stranger, spouse, sibling, worker, or even someone else's kid.

INTERACTIONS

First, we will discuss various different interactions your loved one my have with other residents, then we will cover potential experiences you may encounter with residents.

As Alzheimer's advances and memories fade, one of the strongest emotions remaining is that of a sense of belonging. Your family member needs others – it's human nature to desire friendships and confidants regardless of age or condition. And so, to make sense of the world, your loved one may develop friendships with other residents that may, to the residents, seem as natural as daylight and water.

In similar fashion, sometimes romances blossom as well. A fellow resident may remind them of someone from a younger time, prompting "rekindling" of a relationship. This may be difficult for you, as a family member, to see your loved one developing a relationship with someone who, if absent Alzheimer's, may never have chosen.

THE ALZHEIMER'S PLAN

As difficult as it may be, two major considerations must take place in these situations: your loved one's safety, and your loved one's happiness. When considering safety, you (and the staff) must consider the level of dementia residents have; if they are at similar levels, this minimizes the likelihood of coercion or power struggles. Likewise, another safety consideration involves diligent monitoring for continued consent. If, at any time, one or the other resident demonstrates a desire to not be with the other (such as pushing away, saying "no" or similar), then the two should be gently separated and redirected elsewhere.

Happiness, however short-lived, should be embraced by staff, families, and the residents. If your loved one seems content sitting on the couch with another resident and holding hands, or even believing she is married to the other person, why should we bring them to our reality? A person living with Alzheimer's cannot re-enter our reality, and so they create their own.

> *Mr. Johnson has Alzheimer's disease and lives in a secure memory care unit; his wife of over 40 years lives in the family home. Mr. Johnson no longer recognizes his wife. He has sparked up an interest in Helen, another resident. He and she sit together for meals and can be found holding hands while watching TV together in the evenings. He calls Helen "her" or "she" and Helen refers Mr. Johnson as "Bill" (that is not his name). The nurse has assessed both residents and determined that they are both at about the same level of memory loss. Discussions have taken place with Mr. Johnson's wife as well as Helen's son; both have come to an understand that their happiness, however short-lived, is of primary importance. Staff has been instructed to report to the nurse any time Mr. Johnson and Helen are not getting along, and they will engage each resident in other activities if/when this occurs.*

On the opposite end of the spectrum, there may be residents who demonstrate challenging behaviors that may be concerning to you and/or your loved one.

Some examples include residents who:

- <u>take items from others</u> (most likely they believe those items are theirs to begin with),
- <u>wander into others' rooms</u> (again, they may be looking for something/someone/someplace and have lost the ability to appreciate boundaries),
- <u>cuss or speak gibberish</u> (this occurs when the verbal aspect of the brain is damaged by disease),
- <u>hit out</u> (this may occur in Alzheimer's disease and other dementia-related conditions, as well as a result of perceived personal threat).

A diligent and attentive staff that is trained in the nuances of dementia care will be able to intervene appropriately in many of these situations. At times, however, incidents happen that otherwise could not be prevented. Prepare for these times, as dementia has unpredictable symptom that cannot always be harnessed.

As a doting family member, you will likely experience another residents' company while visiting your loved one. It's likely best to "go with the flow" in these situations, particularly when residents misinterpret the relationship you have (or don't have!) with them. Another resident may believe you to be their family member or friend. Avoiding details of a perceived relationship, while focusing on the person's feelings, may be the best way to handle these types of situations.

Prohibitions by Federal, State, Local and/or Assisted Living Corps

Using Washington State as an example, the following applies:

The Washington State prohibits giving gifts to caregivers, and no fraternization. The logic is that the caregiver may give favorable treatment to the loved one of a person or family that gives gifts to the loved one or fraternizes with a friend of family member of the loved one.

The corporation owning several assisted living facilities prohibits:

- tube feeding
- injections
- lifting downed residents
- As Director Wellness and/or LPN: Cannot delegate responsibilities
- Assisted living corporations each have own policies and procedures
- No delegations and not even as RN
- Can't draw blood
- No urinalysis
- No tubes feeding

Caregivers: Same as Director in prohibitions

- Can't clip nails
- Can't do wound care - band-aid yes but if falls off, can't replace
- Can't pick up fallen resident
- Can't break skin of resident, e.g. no injections
- Can't put meds in resident's mouth
- (Above applies to all staff)

Other Places:

- Have different policies, e.g.
- RN can delegate
- Two people can lift resident
- Can cut diabetic toenails

SUMMARY

Regarding other residents living in the same facility as your loved one, if you are concerned about upsetting behaviors, promptly speak with someone in management and follow up to ensure follow-through. Your goals are the same as your loved one's: safety and happiness. Once those two are achieved, it is important to maintain them throughout your loved one's stay.

NOTE: In Attachment 2, we've provided some stories about residents in assisted-living, adult daycare, and home care facilities. We've done so to recognize their lives and share their stories. When you visit your loved one in such facilities, and visit with some of the residents, you'll be amazed as to their lives and better appreciate what a resident afflicted with this disease has gone through, is going through.

SECTION 4

FROM THE BEGINNING TO THE END

This section is divided into four chapters:

Chapter 4-1: <u>Legal Documents</u> discusses legal documents that will be necessary in order to ensure your loved one receives the care and services needed.

Chapter 4-2: <u>The Beginning of the End</u> talks about indicators of what to expect as your loved one nears the end of life.

Chapter 4-3: <u>The Final Journey — Hospice</u> defines and talks about the Hospice process.

Chapter 4-4: <u>When the End Comes</u> talks about being prepared and steps to take before and after your loved one's life ends.

CHAPTER 4-1

LEGAL DOCUMENTS

Of all the things I miss, I miss my mind the most.
— Unknown

INTRODUCTION

This chapter discusses legal documents that will be necessary in order to ensure your loved one receives the care and services needed. Likewise, considerations regarding financial matters are addressed in order to ensure that your loved one's money is protected.

Old Age or Alzheimer's?

We cannot stress this enough: there is often a fine line between "old age" and Alzheimer's disease. Your loved one may show signs of Alzheimer's that you may mistake for old age and vice versa.

If your loved one lives with you, the changes may happen gradually. That being the case, you may not notice what is happening at first and think these behaviors or memory lapses are simply a normal part of aging, and your loved one will "snap out of it.' If your loved one does not live with you and you don't see them often, the deterioration of their thoughts and actions may be more obvious.

Before your loved one deteriorates further, legal documents should be completed, reviewed, and/or updated. These documents are absolutely necessary because as Alzheimer's progresses, your loved one will be cognitively unable to make even basic life decisions, let alone major decisions such as treatment and living options. Without these documents, medical treatment may be delayed or worse, refused by your loved one altogether.

At a minimum, the legal documents necessary to ensure your loved one gets care and services needed include durable power of attorney for finances and health care (or alternatively, a guardianship), a will, and legal forms instructing CPR or no CPR (forms vary by state). In addition, it's a good idea to review health insurance policies, at least annually at the time of open enrollment, to ensure active and appropriate coverage. Let's break these down, one-by-one, and discuss individually.

Power of Attorney

A power of attorney will probably be the most vital and important document in this type of situation, as it will be needed as the loved one's cognitive functioning deteriorates. This type of document must be initiated and signed by the person, so their level of Alzheimer's cannot be too far advanced or they will be unable to initiate such a task. Ideally this type of document is completed well before a person has any signs of memory loss; however, many people don't think that far ahead.

There are different types of powers of attorney, and your loved one needs two – healthcare and finances. Both need to be deemed "durable," meaning that the instructions within the document remain intact should the person become unable to make decisions on their own. These two types of durable powers of attorney (DPOA) may be separate documents, or combined into one.

DPOA are necessary in the event a person with Alzheimer's needs someone else to make decisions on their behalf, whether those decisions are financial or healthcare related. The person or people appointed as DPOAs must take this role seriously; the purpose of a DPOA should never be viewed as "all encompassing," meaning the DPOA does not make every decision for their loved one, nor do they make decisions based on THEIR values, but rather the values of the person living with Alzheimer's disease.

THE ALZHEIMER'S PLAN

DPOA does not legally go into effect until the person is no longer able to make decisions. Legally speaking, the person must be deemed incompetent. Of course when it comes to Alzheimer's disease, the lines are often blurred. If your loved one lives in a facility, the staff there will often look to you for major decisions (such as treatment plans and medical appointments) and payment for care, but continue encouraging the resident to make day-to-day decisions such as what clothes to wear and what activities to participate in. This is an ideal situation, where the big decisions are made by the DPOA, who is viewed as the "surrogate" decision maker and makes decisions as they believe their loved one would have made them, and the small decisions are made by the person living with Alzheimer's.

Sometimes people holding the DPOA papers view them through a skewed lens, and want to make decisions for the person that is beyond their scope as a DPOA. A great example, particularly in long-term care settings, is that of relationships. People living with Alzheimer's may find romance with another resident living in the same setting. DPOAs might feel as if they have the right to tell the staff to separate the two residents, as they may not approve of the relationship. This is outside the scope of a DPOA's realm of "healthcare" or "finance" decision maker. Likewise, the loved one's choice to be happy (in this case, by being in a relationship) should fall into the DPOA's two primary goals for the person: safety and happiness. Staff at the facility, of course, must evaluate the safety of the relationship and ensure continued safety of both parties.

Once you've obtained DPOA for your loved one (or someone else in your family has done so), it is important to keep a copy of these documents somewhere safe, and share a copy with various persons including your loved one's doctor (so that you may actively participate in medical appointments and make decisions regarding treatment plans), financial institution (so that you may get your name on your loved one's accounts and pay bills on their behalf), and long-term care setting (so you may access the person's health records, make necessary decisions on their behalf, and pay their bills).

Often the financial DPOA goes into effect before the healthcare portion does. As people with Alzheimer's progress into the illness, they tend to make poor decisions regarding money – they may inadvertently write checks from defunct accounts, transfer money to scam artists preying on the elderly, or forget to pay bills altogether. It is imperative, then, that you take note signs that indicate that your loved one is needing assistance in managing finances.

> *One woman with the onset of Alzheimer's was visited by a loved one who found that there were envelopes stamped "Overdue." When inquiring, the woman didn't even know it, as her mail had stacked up. The son told his mom that he could take the bill-paying burden off her hands. She consented. He went through her mail and files and found she had three checking accounts (two of which were empty), an empty deposit box, two CDs, and two credit cards — all at the same financial institution. He talked to her about simplifying her finances for which she was grateful.*
>
> *He took her to the bank, eliminated all but the current checking account, cancelled her safe deposit box and all credit cards. The CDs were cashed in and the funds deposited in the checking account. Of course, one can say that it may be a good investment to keep the CDs to get the interest; that is of course an individual decision and should be based on individual need. All of her financial documents were taken by her son and the loved one's address changed to his so he could manage her money for her. She never missed a thing.*

Guardianship

Occasionally, people living with Alzheimer's don't complete the DPOA paperwork before their cognitive decline is so significant that they are no longer able to make the decision on who will become their DPOA. In this case, guardianship may be in order. This type of document must be completed in the court system and the legal guardian is appointed by a judge. It is a slow and deliberate process that is uncomfortable for the person living with Alzheimer's and the person seeking guardianship, as you may find yourself telling a judge, in the presence of your loved one, how his poor memory has negatively impacted their life and ability to make decisions for themselves.

Guardians are reviewed by the judge on a routine basis (oftentimes yearly) to ensure that the guardianship is still warranted and the process is going smoothly. Generally, a guardianship affords the appointed guardian all decision-making for the person, effective immediately. Guardianships may be limited in scope, depending on the judge's final ruling; for example, the guardian may have the power to pay the person's bills, make decisions on care-related issues and living situations, but not make day-to-day decisions for the person such as clothing choices and activities.

> *Matt was a 64-year-old man who moved into an assisted-living apartment two years ago with his wife; she needed care from staff and he was deemed "independent," not receiving any care or services from the staff there. His wife died, and he continued living in the apartment, paying rent and eating meals there but not getting any care. A few months after his wife died, Matt's behavior changed significantly, where he was found to be physically and verbally threatening other male residents in the building for "looking at my woman" (a female resident unknown to him). A few times he left the building to "go for a drive" and was brought back, days later, by police – he was disheveled and unaware of where he went or how he got there. It was apparent to the staff that Matt had some sort of memory loss, and needed some medical attention. When the nurse offered to make Matt a medical appointment to get evaluated by a doctor, he became irate, screaming at the nurse and physically pushing her against a closed door. Over the coming months, Matt stopped showering or changing his clothes, and became paranoid of the housekeeping staff, eventually refusing their entry to his apartment for routine cleaning. Efforts to contact any loved ones for Matt proved fruitless; his one son lived in Germany and did not want anything to do with him. Knowing that Matt needed care, and ideally in a secure memory care unit where he wouldn't get lost, the administrator at the building where he lived requested Adult Protective Services to seek guardianship proceedings with the court on Matt's behalf. Over a year later, well into Matt's cognitive decline, a judge granted guardianship to a professional guardian company, whose employee promptly moved Matt to a memory care unit for secure care.*

Will

No matter the size (or lack thereof) of one's estate, everyone needs a will. This document outlines final wishes for personal effects, as well as funeral arrangements and disposition of remains. Ideally an executor is also named; this trusted person is responsible for carrying out the instructions in the will that may include distributing money and/or possessions and paying off any debts using the deceased person's funds. Without a will, the state is tasked with deciding who gets what; this impersonal and often drawn-out process delays final wishes and may promote hard feelings amongst family members and friends.

CPR/No CPR Instructions

All states have some sort of legal document regarding wishes should a person go into cardiac and respiratory arrest. In Washington State, there is a form entitled, "Physician's Orders for Life Sustaining Treatment" (POLST). This document not only instructs on whether or not to start CPR, but also guides emergency personnel and hospital staff on the extent of treatment the person chooses. This includes levels of treatment (limited interventions, or comfort-measures only), whether the person prefers to have artificial nutrition/hydration by tube/IVs or not, whether they would like to be transferred to the hospital or not, etc. While these types of legal documents are not required by any setting, it is helpful to identify the extent of care a person may want if one's life is near the end.

This type of document, however called in your state, requires much conversation and consideration. It's designed as a "look to the future" and helps to guide any outside medical professional (including care staff should your loved one move to a facility) in fulfilling wishes.

Should the person living with Alzheimer's be unable to initiate this form due to advanced memory loss, an appointed DPOA or guardian may be able to do so on their behalf; check with your state's laws on this (or, alternatively check with your loved one's doctor).

Many people opt to not complete such a form; after all, it is optional. Know that not completing a form such as this is indeed a decision – without it, full care will be rendered. You may find your loved one, without such a valid document, in the ICU for full treatment or even on life support, even though you know in your heart they never would have chosen this for themselves. It is best, then, to make a decision and complete this document before such a situation arises.

Insurance

October typically starts the health insurance mailings; your loved one may be bombarded with solicitations from health insurance companies as decisions must be made on coverage for the coming year. Your loved one probably needs assistance with this decision, and help completing the necessary documents. Many insurance companies require online applications; this may prove difficult for the person living with Alzheimer's disease. Similar to you choosing your own health insurance, you'll need to consider each company for your loved one to determine if their doctor and preferred hospital is covered, and if the pharmacy of choice is also covered.

For people with long-term care insurance, it's imperative that the required paperwork is completed promptly and returned to the company in order to maintain coverage. Your loved one may be unable to complete this task, and need your assistance. If your loved one lives in a long-term care facility, the staff there may complete the required forms on your loved one's behalf.

> *Clarence has early-onset Alzheimer's disease and lives in an assisted-living community; his care is paid for through a long term care insurance policy. The insurance company sends Clarence a report once a month, that must be completed and returned in order to maintain coverage. He usually gives the form to his son or to the assisted-living nurse to complete and return. In November, he received a document in the mail but forgot to give it to someone to complete. The insurance company sent him a reminder notice in December; he did not respond to that either. In January he was notified that this long-term care insurance policy had been cancelled; his son must now pay for Clarence's stay at the assisted-living facility.*

Other Documents

Various states have other documents, all optional, that may "paint a picture" for what a person living with Alzheimer's wants for the end of life care. Some long-term care facilities incorporate an "advance care plan" that outlines the person's wishes as they near the end of life. One such document is entitled, "Five Wishes"[39] and serves as a guide for families, medical professionals, caregivers, and friends and ensures that everyone knows the person's final wishes as death nears. This document is written in an easy-to-read format and is portable – it can be incorporated no matter where the person lives or seeks care.

> *Vicki Anensen-McNealley's grandfather had a Five Wishes completed before he was stricken with Lewy Body dementia. The assisted living where he lived (and died), kept a copy and when he entered hospice care there, the staff brought her a copy as a reminder of what Grandpa wanted in his last days. As silly as it may seem, he indicated that he wanted Johnny Cash songs playing softly in his room, and ice cream available whenever he indicated that he was hungry. He wanted to be dressed in his Seahawks jersey and hat when he died.*

SUMMARY

It is imperative that family members and friends know the signs of both old age and Alzheimer's (and other forms of dementia). As your loved one gets older, begin to look for signs of mental deterioration. Prepare a plan of action that covers both healthcare and financial issues such as a will, durable power of attorney, reviews of the loved one's finances, and insurance policies.

---

[39] fivewishes.org

CHAPTER 4-2

THE BEGINNING OF THE END

Our dead are never dead to us, until we have forgotten them.[40]
— George Eliot

INTRODUCTION

This chapter talks about indicators of what to expect as your loved one nears the end of life.

As the Disease Worsens

As a relative of the person diagnosed with Alzheimer's disease, you may want to care for that person until the end. That is a very compassionate and admirable goal. Our discussions so far indicate in general what you can possibly expect from your loved one. However, those have been through it can attest to totally draining, emotional experience one must endure on a 24/7 basis. (See Chapter 2-3 for deciding what to do about your loved one).

The Alzheimer's sufferer must have a responsible adult in attendance 24/7. If you think that you can handle it, maybe you can. At times you may think that your teenage daughter or son can watch that person, and maybe they can. The phone may ring, you or one of your kids are busy doing something, get pre-occupied, turn your back for only a few moments, return and them find gone.

---
[40] wisdomquotes.com

You search the house, yard, neighborhood and can't find them. The person is gone. What happens if you find the car is also gone, that the person has not only left the house but is now driving somewhere? Think about it. How would you feel? Have you ever watched the news when they report that an elderly person is missing and ask the viewers to call the police if they see the person? This is the type of situation that all would prefer to avoid and yet it is a reality for many family caregivers.

As your loved one's Alzheimer's disease worsens (and it will as currently there is not a cure) you must be prepared for their downhill slide. You may have known this person as a wonderful, caring individual; however, their personality may change to the point of becoming almost another person over time.

> *One woman in the later stages of Alzheimer's was in her room, when her son entered. She did not recognize him and began screaming for this "stranger" to get out and calling for help. To her, her son was a complete stranger.*

Some of the indicators[41] are noted below. And as stated on the website: "…The stages don't always fall into neat boxes, and the symptoms may vary…"

- Recognize faces but forget names
- Mistake a person for someone else
- Delusions such as need to go to work but retired
- Needs help going to the bathroom
- Basic abilities may fade such as eating, walking, sitting
- Need help holding or using eating utensils
- Needs to be fed, maybe only soft foods such as apple sauce or pudding
- Be sure that the person drinks plenty of liquids, e.g. water, maybe juices (depending on the person's required diet, e. g. diabetic)

---

[41] See alz.org and webbed.com/guide/alzheimers-disease-stages

Why? They may no longer know when they are in need of water or even thirsty
- Helping him/her use a spoon.

> *"Arizona woman, 92, shot, killed son who tried putting her in assisted living, cops say"*[42]

Keeping Your Loved One's Mind Active

Whenever you are with your loved one, you must always be upbeat with a smile at all times. Keeping a positive attitude, no matter how bad a day you had, how stressed and depressed you may be, will help those there and especially your loved one feel better. This does not always work but do it anyway. Remember that there will probably be other residents around who may hear or see your actions.

If your loved one is in a care facility, before going to see your loved one, prepare things to talk about and do. For example:

- If they had enjoyed working crossword puzzles, bring in a book to them that is very basic. Then work them together. If they get frustrated, you can mitigate that by explaining that you brought them in as you like to do them but have a difficult time and could use his help.
- Bring in a list of the best 100, one-liner jokes to share and ask your loved one to alternate reading them with you. If your loved one can no longer read (this may be a symptom of Alzheimer's), then you can read them out. Talk about each joke, and laugh together.

---

[42] Fox News, 7/3/18

DR. KOVACICH AND DR. ANENSEN-MCNEALLEY

> *Dr. Kovacich did that with his mother in an assisted-living facility. After reading a few jokes, she asked for some coffee. He brought it back and they continued; however, she started reading from the top as she didn't remember the ones read, even though less than five minutes had passed.*

- Another good way to exercise your loved one's hand, eye, and brain coordination is to bring in a simple drawing book and pencils, paper. Then each draw the same object. Don't be embarrassed if your loved one does a better drawing than you. Regardless, tell them they did great and better than yours. That helps them have confidence in themselves and a positive attitude.
- Download some of your loved one's favorite music and listen to it together. Maybe ask them the words to that song. Don't be surprised if they remember from the good ol' days and start to sing.
- Another thing to do is put a booklet together of photos of her family and friends. Label the name and relationship of each. Periodically go through it together reinforcing the relationship. Avoid "quizzing" your loved one about pictures. For example, you may find yourself saying something like, "Who is in this picture?" With Alzheimer's disease, your loved one may not remember who the person is and become frustrated. Instead, ask general questions or make general comments like, "I love that lady's dress, What do you think about it?" Or "Tell me what you think of when you see this picture." This allows the person to talk about whatever comes to mind, instead of trying to remember specifics.
- Set up a Skype or FaceTime with family members to help re-enforce the relationships. Try your best not to get upset if your loved one does not remember the person on the call make it a light-hearted visit and simply enjoy each other's company.
- Think of their life and what they enjoyed. If they love to fish for example, bring your computer and play a fishing video; bring

photos of them fishing; talk about fishing. If they love old movies, musicals, etc., play them; talk about them.
- What other activities can you think of to do with your loved one?

SUMMARY

After familiarizing yourself with indications of old age, Alzheimer's and other forms of Dementia, look for such indicators in your love one. When indicators are there, of course must be confirmed by medical professionals.

When confirmed, establish a plan of action and implement as necessary. Always keep a positive attitude around your loved one. Find challenges for them during your visits. Never contradict them as they will usually become confused or freeze.

CHAPTER 4-3

THE FINAL JOURNEY — HOSPICE

If we lose love and self-respect for each other, this is how we finally die.[43]
— Maya Angelou

INTRODUCTION

This chapter discusses the definition of hospice and its uses for people with Alzheimer's.

Hospice

What is Hospice? Hospice care is a type of care and philosophy of care that focuses on the palliation of a chronically ill, terminally ill or seriously ill patient's pain and symptoms, and attending to their emotional and spiritual needs. In Western society, the concept of hospice has been evolving in Europe since the 11th century.[44]

Then, and for centuries thereafter in Roman Catholic tradition, hospices were places of hospitality for the sick, wounded, or dying, as well as those for travelers and pilgrims. The modern concept of hospice includes palliative care for the incurably ill given in such institutions as hospitals or nursing homes, but also care provided to those who would rather spend

---

[43] brainyquotes.com
[44] Wikipedia

their last months and days of life in their own homes. The first modern hospice care was created by Cicely Saunders in 1967.

There are differences between what has been deemed "palliative care" and "hospice care" in westernized medicine. Not to confuse the two, families of people living with Alzheimer's may seek out palliative care and ultimately end with hospice care as the person's condition declines significantly as to warrant end-of-life care.

Palliative care is offered by many insurance companies and focuses on effective management of serious illnesses, including Alzheimer's, with a quality of life as the prime focus. Its holistic approach to disease management can be accessed well before the dying process begins, and is often a good stepping stone prior to hospice care.[45]

In the United States the term is largely defined by the practices of the Medicare system and other health insurance providers, which make hospice care available, either in an in-patient facility or at the patient's home, to patients with a terminal prognosis who are medically certified at hospice onset to have less than six months to live. According to the National Hospice and Palliative Care Organization (NHPCO) 2012 report on facts and figures of hospice care, 66.4% received care in their place of residence and 26.1% in a Hospice inpatient facility.[1][2] In the late 1970s the U.S. government began to view hospice care as a humane care option for the terminally ill.

In 1982 Congress initiated the creation of the Medicare Hospice Benefit which became permanent in 1986. In 1993, President Clinton installed hospice as a guaranteed benefit and an accepted component of health care provisions.

Outside the United States, the term *hospice* tends to be primarily associated with the particular buildings or institutions that specialize in such care (although so-called "hospice at home" services may also be available).

Such institutions outside the US may similarly provide care mostly in an end-of-life setting, but they may also be available for patients with other specific care needs. Hospice care also involves assistance for patients'

---

[45] https://www.hospicealliance.org/what-is-hospice/hospice-vs-palliative-care/

families to help them cope with what is happening and provide care and support to keep the patient at home.[46]

Hospice care may be provided at home or if your loved one lives in a facility, hospice care can be offered there too.

What Exactly Do Hospice Agencies Provide?

In general, hospices provide a variety of services that supplement or replace services provided by the facility or home where your loved one lives, whether that be at home under your care, home caregiver or in any of the facilities described earlier.

These services generally begin when the person diagnosed with Alzheimer's by a medical provider, such as a doctor, to be within six months of death. The health conditions that determine hospice eligibility are expansive and could focus directly on Alzheimer's disease, or the person's other chronic medical disease such as diabetes, congestive heart failure, or emphysema to name just a few. At this time, the loved one might be bed-ridden and a hospital-style bed is provided to replace the normal bed. This provides some degree of safety with the bed rails (if allowed in the care setting); as well as a degree of convenience for the caregiver and also your loved one as the elevation of the bed can be adjusted.

Hospice services generally provide the following:

- Experienced specialists in dealing with the needs of the Alzheimer's victim
- Skilled nurse to manage health-related symptoms and address pain management
- Bathe the loved one in a compassionate and caring way
- Religious or other type of spiritual person for them to talk to, but only if requested. This service or any other for that matter is up to you and if they still are having the ability to make such decisions, the loved one
- A "conversationalist" to just sit and talk to your loved one

---

[46] Ibid

- Counseling on the disease process and what to expect as your loved one's health declines
- Possibly help with medication administration
- Other support that is requested by or recommended for the patient or family.

Hospice support does not end at the death of your loved one. Hospice agencies also provide counseling groups that you can join that deal with bereavement; counseling is available before as well as after the death of your loved one.

The cost of hospice care is generally provided through Medicare and other insurances, depending on the loved one's circumstances. It is also an institution that normally is one that you can give to as a charity with its charitable contributions being tax deductible — at least at the time of this writing. You should of course verify this with your tax professional and the charity status of the hospice.

When and who calls in the hospice support professionals of course varies between care agencies and you. Once it has been determined that your loved one is in the final stages of life, you should talk to the staff at the facility where your loved one lives or, if at home, directly with your loved one's primary care doctor. For hospice staff to provide care, a primary care provider must recommend an evaluation; the hospice agency nurse will determine whether your loved one qualifies, based on a thorough physical assessment as well as reviewing the history of your loved one's condition and overall health decline. The legal decision maker for your loved one, whomever that is, can choose which hospice agency will care for them; the nearest one can easily be found by talking to your doctor, care facility staff, home caregiver or a simple online search using one of the many search engines available.

SUMMARY

A hospice agency is a valuable support entity that offers many services during the approximately last six months of an Alzheimer's person's life. It not only helps your loved one, but also your loved one's family before and even after they die.

CHAPTER 4-4

WHEN THE END COMES

Life should not be a journey to the grave with the intention of arriving safely in a pretty and well preserved body, but rather to skid in broadside in a cloud of smoke, thoroughly used up, totally worn out, and loudly proclaiming "Wow! What a ride!
— Hunter S. Thompson[47]

INTRODUCTION

In this chapter, we discuss when the end comes, talk about being prepared and steps to take before and after your loved one's life ends. If you followed our advice earlier in this book, you will be at least somewhat prepared for the death of your loved one.

Being Prepared

You should always be prepared for the end of your loved one's life. Some people go into denial and figure their loved one has years to live. Yes, maybe but much of that depends on the stage of Alzheimer's they are going through, other health issues, age, etc.

---

[47] wisdomquotes.com

> *Remember, currently there is not a cure for Alzheimer's disease and your loved one's health won't get better or be stagnant. The person's mental state will continue to deteriorate over time. New drugs may help, may delay the inevitable, but so far the disease cannot be cured, nor can the person, reverse their mental state — at least not yet.*

A Current Will

As previously stated, you should begin early to prepare for the death of your loved one. When it comes to a Will, this is very important. If not completed by your loved one before or in initial stage of Alzheimer's, then if relatives do not agree on the distribution of the estate, one of the challenges may be that the person was not in sound mind when the will was made. Besides, everyone should always have a Will. As the saying goes: "Never know when you're goin' to go!"

Hospice Support

As the end nears, one should coordinate with the victims' healthcare providers as to hospice care. They should be simple if you in fact already coordinated with your local Hospice agency in preparation.

Hospice staff can provide services as stated earlier and thus, take some of the burden off of you. Remember that, agency-dependent — hospice provides services during the last six months of your loved one's life. That fact alone should get you and your relatives to understand that the end is near.

THE ALZHEIMER'S PLAN

> *When Dr Kovacich's mother was diagnosed, she agreed to have her Will updated, and cremation and burial services arranged and paid for. When she died, the assisted-living staff where she resided called the burial agency (all that was pre-arranged). They picked up her body, cremated her and had her ashes transported to the grave site she already had paid for, next to her husband. The burial was in another state and her ashes were transported by UPS and again, due to prior arrangement, the burial staff received the ashes and buried her ashes based on prior coordination. Per her direction, that is what she wanted done and no services. Having such arrangements made early is convenient and relieves the loved one's family of such tasks at a time when grieving and stress are dominant.*

Notifications

Don't forget to also have a list of names and address, phone numbers, email addresses of those you want to notify upon the death of your loved one. If prepared in advance, it's just a matter of hitting the send button and group mailing goes out. For those without email address, a simple phone call should suffice.

> *The amount of public notice of your loved one's death you do with all good intentions may also be used by those wanting to steal the loved one's identity for fraudulent purposes. Furthermore, if a public funeral or notice of a private funeral, burglars may be reading the obituaries and while you are at the funeral home or burial site, your home may be robbed. Keep that in mind when having the loved one's death published.*

Burial/Cremation and Funeral Plans

Other matters that you should talk about in advance with your loved one are burial/cremation, and funeral plans. These rather unpleasant conversations require a level of detail that forces decisions: where will the body be taken immediately after death; which company will manage the care of the body? Will there be a burial or a cremation?

Similar to burial/cremation plans, a funeral or celebration of life should also be discussed. This ritual is likely to be unique based on your family's past history and perhaps religious or spiritual habits. By planning ahead, this event can come together with minimal effort once your loved one has passed away.

These are distasteful matters to discuss and many avoid any conversations about death and all that it entails. Like the Will, the sooner you can sit down with your loved one who has been diagnosed with Alzheimer's and plan for their end of life, the better.

One difficulty you may face is that they may not want to talk about it even before an Alzheimer's diagnosis and probably less so after the diagnosis. Some people living with Alzheimer's may be in denial and won't discuss end-of-life matters at all; actually, this may apply to you and other relatives as well. Some may want to put it off until later; unfortunately, later comes and these decisions can be tremendously difficult to make when also dealing with emotions surrounding a person's death. However you feel about the topic, the conversations still must take place.

As part of that "final six months" you should finalize preparations for cremation and/or burial services by contacting your burial services agency to ensure all plans are in place. That way, when your loved one dies, you or maybe your assisted-living provider, of course depending on your loved one's living location and such, will be more prepared. That way, one phone call to the burial agency will set in motion all that needs to be done.

There are many decisions that must accompany this one major choice — consideration for a burial casket, clothing for the loved one, whether there will be a viewing of the body, where the burial plot will be located, and details associated with a headstone/plot marker. For a cremation, a decision must be made on what type of urn to choose and where the ashes are to be distributed or stored.

As you can imagine, circumstances surrounding the death of a loved one is stressful enough. The ore prepare, the less stress should be felt.

Death Certificates

You will probably need multiple copies of the death certificate. This should be arranged in advance with the burial agency who will provide them. Always get extra certificates. It is easier to obtain several at one time than to request additional copies later. Depending on relationships and feelings of relatives, each relative may want a copy. This is especially needed for those, who like so many these days, do research and write the family history.

Copies of the death certificate must be sent to the Social Security Office and other agencies who were providing retirement payments to the loved one. Also, you'll need to send a copy to companies such as health insurance and life insurance. If you have a joint account with the deceased and use their retirement money after death for other than to clear up your joint financial obligations, your actions made constitute a fraud and consequently, you can be prosecuted for fraud. It is essential then to use your loved one's money solely to finalize their last bills. Check with your lawyer as the debts of your loved one may not be your responsibility when they are deceased.

You should have maintained a list of names and addresses of institutions who are providing income to your loved one — including a form letter. When the time comes, place the modified form letters with attached death certificate and just mail them in.

Finally, there's the reading of the will and associated lawyer fees for that, if required. As the executor of your loved one's estate, and assuming you have a joint checking and/or other accounts, it is easier to wait until all of the loved one's financial obligations are paid before closing those accounts.

If closed too early, things get more complicated in dealing with the related financial institutions who probably will require a copy of the will, death certificate and after it is reviewed and, hopefully approved, by the financial institution's legal bureaucracy, many months may have gone by. You may even have to go to court to clear legal hurdles before you can properly execute your deceased directions.

SUMMARY

When the end comes, you should already have been prepared: a plan or plans in place that just have to be implemented. Factors to consider are support from professionals, execution of the will, power of attorney, final financial payments for any outstanding bill, and burial arrangements.

SECTION 5

LESSONS LEARNED AND THE FUTURE

This section, divided into three chapters, provides a summary of dealing with your loved ones Alzheimer's diagnosed.

Chapter 5-1: Summary of Alzheimer's provides an overview of the latest developments related to this disease, and status of research.

Chapter 5-2: The Future of Dementia-Alzheimer's Disease and Cures provides a look beyond current research and the possibilities of eliminating this disease.

Chapter 5-3: What You Can Do to Help talks about things you can do to help fight Alzheimer's.

CHAPTER 5-1

SUMMARY OF ALZHEIMER'S

"After you find out all the things that can go wrong, your life becomes less about living and more about waiting."
— Chuck Palahniuk[48]

INTRODUCTION

Over the course of our experiences with this disease, there have been various theories on the cause of Alzheimer's. We discuss some of them; as well as other Alzheimer's-related issues.

Alzheimer's Theories

Some Alzheimer's theories:

- For a long time, we believed that people living with the disease simply did not work their brains often enough; for that reason, we encouraged crossword puzzles and other brain games to keep the mind active in the hopes of staving off the disease. This theory often did not hold true simply because in care facilities, staff experienced patients who were scientists, priests, doctors, mechanical engineers, and other professions that required much brain work and yet these people experienced similar decline to the

---

[48] goodreads.com

housewives, plumbers, and millworkers who were being cared for. The level of education and brain use didn't seem to make much difference to the plaques and tangles attacking either of these classes of residents.[49]

- Another theory at one time encompassed the notion of a build-up of aluminum in the brain, as autopsies have shown this in brains of people who died because of Alzheimer's. This prompted healthy people to change their antiperspirant usage to minimize the threat of aluminum absorption; many people too stopped purchasing food products stored in aluminum and changed their pots and pans to stainless steel or cast iron.
- A theory exists today focuses on lack of sleep. In people who do not get enough deep sleep, the brain cannot rid itself of contaminates and rejuvenate. Chronic insomnia may lead to cognitive decline and eventual permanent memory loss.

The fact is, you must do your own research and come to your own conclusions on what is best for you and your loved one. The theories mentioned continue to have supporters and deniers, as often occurs with such issues. Further research must be done in order to determine causes so that effective prevention and treatment options can be discovered.

Some think that Alzheimer's (and other dementia-related diseases) are related to inflammation and the body's response to it (immune system) in some way. If research has found that diabetes, high cholesterol, hypertension, obesity, and other chronic conditions are related to the body's overall inflammation, why would it not be the same with Alzheimer's? After all, the brain is connected to the body.

To further that theory, research is linking chronic body inflammation to sugar intake. Added sugar seems to be lurking in the most common of foods – fast food, snacks, drinks, breads, soups, cereals, spaghetti sauce,

---

[49] In prior chapters of this book, we discussed some of the things to do with your loved ones that included cross-worded puzzles, and other exercises to keep the brain active. We believe that is helping keep the loved one active, challenging them, entertaining them, etc. However, it will not help the loved one win her battle with Alzheimer's.

ketchup, lunchmeat...it's literally everywhere. It's hidden in various forms – it can be called sugar, sucrose, glucose, fructose (basically anything ending in an "ose"), high fructose corn syrup, and corn syrup (just to name a few).

Two hundred years ago the average American ate two pounds of sugar a year. Today the average American eats three pounds a week. Combining this fact with the advancement of medical treatments, we've created the perfect storm for chronic diseases to settle into the older American.

There are other conditions and choices that can cause body inflammation besides diet: lack of exercise, stress, pollution, allergies/sensitivities, medications, etc. This demonstrates to the reader that chronic inflammation can be caused by multiple sources, so changing only one aspect of one's overall health will likely not minimize the continuation of then inflammation and the disease will continue to flourish.

Research tends to have two prongs: prevention and treatment. The money for research lies with the latter, as drug companies make more money on treatment options as opposed to prevention or cure. Research on your own, and don't be afraid to venture into the unique or more natural theories.

Learn more about specific types of dementia:

Alzheimer's Association: www.alz.org
Dementia Society of America: https://www.dementiasociety.org/

Learn what it's like to have dementia:

Second Wind Dreams Virtual Dementia Tour: https://www.secondwind.org/

Alzheimer's research centers in the United States

The National Institute on Aging funds thirty Alzheimer's research centers across America; these centers are situated in universities and focus on identification of cause and effective treatment modalities. There may be opportunities to gain excellent Alzheimer's-related information from these centers, receive diagnosis and treatment, participate in a study or clinical trial, and gain access to support groups where other family members are experiencing similar issues with their loved ones. Of these centers, the following list (while not inclusive) includes those with promising outcomes and accessible information.

- University of California, San Francisco -- Memory and Aging Center (https://memory.ucsf.edu/) offers information on trends, research, support, and treatment options. Of note, this center offers routine educational sessions, free of charge, specifically geared towards not only professional caregivers (nurses), but also for family members of loved ones living with Alzheimer's and other dementias.
- University of Washington, Seattle-- Memory and Brain Wellness Center (https://depts.washington.edu/mbwc/) while focused much in research, it also features many opportunities to engage with others in the community. For example, monthly dementia-friendly outings to the Frye Museum and Café as well as Garden Discovery Walks for anyone living with Alzheimer's or other forms of dementia. Their caregivers promoting access to the greater community, where the person living with Alzheimer's enjoying interactions with others who are experiencing similar dementia-related life issues.
- Oregon Health Sciences University (http://www.ohsu.edu/xd/health/services/brain/getting-treatment/diagnosis/alzheimers-aging-dementia/index.cfm) is known for its research on how complementary medicine impacts Alzheimer's and other dementias. Some focuses include yoga, gingko biloba, and placebo effects on memory.
- Wake Forest University, Winston Salem North Carolina (www.wakehealth.edu/alzheimers/) focuses research on preventing cognitive decline in older adults. They host a memory assessment center, and were the first center to develop and implement a research center on Alzheimer's in the African—American population.

- New York University's Alzheimer's Disease Center (www.med.nyu.edu/adc) has been around since 1973 and its research has impacted how health professionals assess and treat Alzheimer's disease today. They are known for the development of various assessment tools used to determine a diagnosis of Alzheimer's disease, as well as the use of MRI to demonstrate how the brain may be impacted by the disease. They not only host research, but also offer a toll-free caregiver helpline.
- The Mayo Clinic (Arizona, Florida, and Minnesota) (https://www.mayo.edu/research/centers-programs/alzheimers-disease-research-center) has a robust Alzheimer's/dementia program that encompasses research, patient and family education (including in-person seminars and online videos), publications, diagnosis and treatment, and caregiver resources.
- University of Kentucky's Alzheimer's Disease Center (http://www.uky.edu/coa/) has been around since 1985 and focuses their research and clinical trials in three main areas – those people who have no memory issues (to determine how or why they might develop cognitive decline), those with current memory issues, and minorities (to investigate how dementing illnesses may affect brains differently). They are also conducting research on the connection between the aging brain in people living with Down's Syndrome, and how that may lead to Alzheimer's. There are several online educational opportunities for family members to learn more about memory loss and Alzheimer's disease.

As mentioned, there are thirty Alzheimer's Disease Research Centers across the nation, all of which have websites and are available as educational, investigative, diagnostic, and treatment resources. More can be found at https://www.nia.nih.gov/health/alzheimers-disease-research-centers.

SUMMARY

There are several theories as to the cause and treatment of Alzheimer's. They include such things as not enough brain exercise, lack of sleep, effect of other diseases, aluminum in the brain, and body inflammation due to sugar intake.

Research primarily focuses on two areas: prevention and treatment are both needed to identify a cure for this disease.

CHAPTER 5-2

THE FUTURE OF DEMENTIA— ALZHEIMER'S DISEASE

Without memory, there is no culture. Without memory, there would be no civilization, no society, no future.
—Elie Wiesel[50]

INTRODUCTION

The thought of Alzheimer's going away anytime soon is, sadly, not one built on optimism. The number of people being diagnosed with a dementia-related condition is soaring by the day, and there is simply not enough money, care settings, nor caregivers to keep up with demand.

National Efforts

A national effort to beat Alzheimer's disease has been in the works since 2015. The most recent update to the plan includes lofty goals including "prevent and effectively treat Alzheimer's Disease and related dementias by 2025" and "expand supports for people with Alzheimer's disease and related dementias and their families."[51] Many states have paid attention to these national efforts and have likewise designed their own. Washington State has a "Dementia Action Collaborative" that includes state leaders

---

[50] brainyquote.com
[51] aspe.hhs.gov/report/national-plan-address-alzheimers-disease-2016-update

as well as advocacy groups and members of the public; these groups have designed a "dementia road map" for families and care partners, along with a "dementia safety kit" for consumers as well.[52] Other states have similar coalitions, with a focus primarily on public education and involvement.[53]

Those families with strong ties are likely to find support amongst themselves and others experiencing similar plights via self-education, informal care, and hospice support.

> *From our personal experience, attempting to care for a loved one alone can drain all aspects of patience, kindness, and positive attitude from one's life; the sense of aloneness may be too much to bear. Finding a balance by asking for help and taking necessary time away for yourself are two factors that will allow you to be a better family caregiver.*

For the rest, without family support, there is an overarching fear amongst healthcare professionals that there simply won't be enough resources to provide adequate care to the growing numbers of people living with this condition. There may not be enough caregivers, not enough long-term care facilities, or not enough money to provide for the care. Without a cure in the foreseeable future, regardless of the national effort's plan to eradicate this disease, the future is dim indeed.

Perhaps a bright note on the horizon is one of prevention. While this idea may not help someone living with Alzheimer's now, it can certainly help the reader of this book. Take your health seriously – eat food (not processed, fast, or "fake" foods) – organic fruits and vegetables, nuts and seeds, and animals/animal products (if you indeed eat meat) that is fed grasses and lives a wholesome life outside of a feedlot.

Move your body every day – walk, jog, swim, ride a bike, do Yoga or Tai Chi or some other form of exercise. Sleep deep and well. Reduce emotional stress – find ways to practice kindness to others and yourself. Be part of the greater community – volunteer about something of which you feel passionate, spend time socializing with friends or family. Use natural

---

[52] dshs.wa.gov/altsa/stakeholders/alzheimers-state-plan
[53] act.alz.org/site/PageNavigator/state_plans.html

skin and body products free of chemicals. Drink plenty of clean water. Stretch your mind by learning something new or trying out a new hobby.

While the future of Alzheimer's may warrant a sense of fear, there are things you can do – for yourself, for your loved ones, for your community. Will this prevent you from developing Alzheimer's? Maybe, maybe not; however, it may delay it. You'll be healthier, maybe fend off diabetes, high blood pressure, high cholesterol, maybe even strokes or heart attacks. In other words, these health-focused efforts certainly can't hurt.[54]

Promising research

There seems to be a new research article published weekly, if not daily, around the globe on Alzheimer's disease and other dementias. These various research findings are promising and demonstrate cutting-edge knowledge that will certainly lend to additional research.

Recently, researchers developed this "brain pacemaker" where electrical wires are inserted into the brains of three patients with Alzheimer's. They wanted to see if the electrical impulses from these wires could help improve cognitive, functional or behavioral abilities in the study participants. They found that the device may slow the declines tied to Alzheimer's disease. They described one patient who wasn't able to prepare meals when she entered the study. After two years, she could prepare meals again, these researchers said. They noted that all three patients showed improvements after treatment…[55]

Scientists have discovered a protein that fuels Alzheimer's disease. This finding has created a shift in how global research has assisted in the fight against Alzheimer's disease, as well as it has opened the door to promising new treatments.[56]

---

[54] At least that is the opinion of the authors and others.
[55] https://www.rxwiki.com/news-article/deep-brain-stimulation-implant-showed-promise-alzheimers-treatment?utm_source=2534731155&utm_medium=newsletter&utm_campaign=RxWiki
[56] https://www.alzinfo.org/research/

Scientists believe that a combination of genetic, lifestyle, and environmental factors influence when Alzheimer's disease begins and how it progresses. [57]

> *Alzheimer's disease—for which there is no effective conventional treatment or cure—affects an estimated 5.4 million Americans and prevalence is projected to triple by 2050.*
>
> *Lifestyle choices such as diet, exercise, and sleep can have a significant impact on your risk. Prevention guidelines are included.*
>
> *Having one or two ApoE4 genes raises your lifetime risk to Alzheim'ers between 30 and 50 percent respectively. Meanwhile, research suggests high-carb diets can increase your risk of dementia by 89 percent, while high-fat diets lower it by 44 percent.*
>
> *Alzheimer's disease is intricately connected to insulin resistance; even mild elevation of blood sugar is associated with an elevated risk for dementia.*[58]

A new study suggests that reduced sleep and poor sleep quality may be linked to increased build-up of beta-amyloid plaques in the brains of older adults - a sign of Alzheimer's disease. This is according to a study published in the journal JAMA Neurology.[59]

Look Out for Unproven Claims about Alzheimer's Treatments

Remember the saying, "if it seems too good to be true, it probably is?" Unfortunately, when faced with a serious health issue, even the most

---

[57] https://www.nia.nih.gov/health/alzheimers/causes
https://www.usatoday.com/story/news/2016/05/26/harvard-researchers-unveil-new-alzheimers-theory/85004894/

[58] https://articles.mercola.com/sites/articles/archive/2018/02/22/link-between-sugar-and-alzheimers.aspx?utm_source=dnl&utm_medium=email&utm_content=art1&utm_campaign=20180222Z1_UCM&et_cid=DM187694&et_rid=222606645

[59] https://www.medicalnewstoday.com/articles/267710.php

rational person can be led to believe implausible claims. Indeed, that's what companies selling fake treatments count on.

One of best ways to protect yourself from fake treatments is to ask whether the claim seems too promising and if it contradicts what you've heard from reputable sources about treatments for Alzheimer's disease. Companies selling unproven Alzheimer's treatments often includes a range of unsupported and expansive claims about the supposed healing powers of their products. These includes statements such as:

- "You can even reverse mental decline associated with dementia or even Alzheimer's in just a week"
- "Clinically shown to help disease of the brain such as Alzheimer's and even dementia"
- "Supplements are used to cure Alzheimer's disease"
- "Can … reduce the risk of Alzheimer's by half"
- "May have a role in preventing the progression of Alzheimer's" and
- "Clinically shown to help disease of the brain such as Alzheimer's and even dementia."

Another red flag is that many of the claims made by these companies about the supposedly curative powers of their products are often not limited to Alzheimer's disease. Consumers should steer clear of products that claim to cure or treat a broad range of unrelated diseases.

For instance, the FDA sent a warning letter to a company selling one product marketed as a dietary supplement that the company claimed that they may help alleviate the neurodegenerative effects of Alzheimer's disease and Parkinson's; that it was successful in "relieving pain associated with rheumatism, back ache and arthritis; can help fight off colds and flu; can help with blood sugar management; and help stabilize high blood pressure; [and] help fight off bacterial infections."[58]

Such claims may or may not be valid. You should do your homework and judge them for yourselves.

DR. KOVACICH AND DR. ANENSEN-MCNEALLEY

SUMMARY

The future is unknown; however, over time scientists seem to research most diseases and find a cure (or, at the very least, excellent treatment options). Lessons and techniques learned fighting other diseases mean that we are not always fighting diseases from scratch. Researchers continue to try innovative methods to better understand the brain and how it works.

By learning more about new research findings and new techniques, as well as experiments being tested, will aid your progress. Also look into trial programs for which your loved one may be qualified to be a candidate.

CHAPTER 5-3

WHAT YOU CAN DO TO HELP

"No one is useless in this world who lightens the burdens of another."
— Charles Dickens

INTRODUCTION

No one can fight this disease alone. It takes all of us working together as a team. Without the team of supporters, caregivers will flounder. That does not mean that everyone needs to become a caregiver; we all have gifts and some people just aren't cut out for caregiving. There are many ways to be helpful, to lessen the burden, to show that you care.

Becoming Involved

Becoming involved in the fight against Alzheimer's is multi-faceted. This involvement is based on your comfort level and sense of obligation, along with your personality. Your help can be small or huge, depending on your desire.

Start Small

Consider taking a class/training on dementia and Alzheimer's. The Alzheimer's Association, Alzheimer's Society, and local assisted living communities may have training at no cost to you. These classes can help you better understand all aspects and types of dementia and how it affects the mind and body. With this understanding, your assistance can flourish.

People living with Alzheimer's may be/are cared for by family caregivers. The life of a family caregiver can be a lonely one. With little support and few resources, it is easy to see how burnout can lead to resentment and ultimately depression in the caregivers. Reaching out to a family caregiver can dramatically impact a life, while positively affecting yours. There are many ways to help, from running errands, delivering a meal, or offering to sit with the person while the family caregiver takes a few precious hours of personal time away. It's likely that you have a neighbor, co-worker, friend, or someone from your church or other associations who is a family caregiver. Reach out today and find out what you can do to lift the burden.

> *Please don't rationalize saying you don't have time. Maybe just an hour a week. How many hours do you spend watching television, on social media, etc. Ok, so some of you don't really have the time due to work, and kids. However, it is interesting just how many people find the time to get involved once their loved one is diagnosed with this condition. If no time, please just add an Alzheimer's-related charity to your charitable contributions.* Your contribution may tip the balance in finding that elusive cure.

Assisted-living communities and nursing homes are always looking for volunteers. By lending even a few hours a month, you can make a difference in the life of a person living with Alzheimer's. You can sit with them, play games with them, read to them, pray or listen to music. You can take them for walks or play Bingo or cards. You can garden with them in the spring. Find out what a resident likes to do and do that. Your time will brighten the day of another. You may have the same interests which would make it even better.

If you're interested in raising money for a cause (and/or you are a natural-born organizer), you can volunteer with the Alzheimer's Walk to Cure Alzheimer's (they also have a Bike to Cure Alzheimer's). This annual nationwide event brings people of all kinds together to raise money to fund dementia research. There are other reputable fundraisers available throughout the country as well; look into local offerings that fit your needs and interests.

SUMMARY

If you have a loved one living with Alzheimer's, of course you're helping in various ways. Be willing and open to others offering their help, and graciously accept in order to avoid your burning out. You should not go it alone – take whatever is offered, however small. Your overall health and inner peace will allow you to better care for your loved one and if they are in a caregiving facility, you'll have less stress and be able to spend happy, quality time with them.

Help comes in many forms. All are important, all are necessary, and all are welcomed.

FINAL QUOTE

As she got old,
She became more forgetful.
She joked, "Of all the things I miss,
I miss my mind the most".
Then gradually, Alzheimer's
Attacked her.
No longer a joke,
Her memory faded away,
As did her life.[60]

---

[60] Taken from the book. Poems of Life: Thoughts of Human Experiences, by Gerald L. Kovacich

APPENDICES

Appendix 1: Alzheimer's Checklist Plan

Appendix 2: More Stories of Dementia—Alzheimer's Sufferers

Appendix 3: Memorial to Helen M. Kovacich

Appendix 4: Thoughts of Grandma and Grandpa Filmore

Appendix 5: About the Authors

APPENDIX 1

ALZHEIMER'S CHECKLIST PLAN

If your loved one is diagnosed with dementia, specifically Alzheimer's for our purposes, you may be scared, stressed and all that coming at you with the knowledge that your loved one will die from this disease, as there isn't a cure. Worse yet, they may be further along than you may think, since for some time you thought it was just old age.

To help you establish a plan for an orderly process and preparation for the deterioration of your loved one's health, get their affairs in order, we offer this overall checklist of "things-to-do". This is not an all-encompassing checklist as each person's circumstances are unique. However, we hope you can use it as a basic guide of the things to consider when dealing with your loved one.

Non-Medical Diagnosis

_____Recent memory loss
_____Language changes
_____Personality changes
_____Delusions (believing things that are not real)
_____Disorientation
_____Sundowning (increased confusion/behaviors late in the afternoon/evening)
_____Wandering
_____Impaired judgment
_____Unable to create and carry out plans
_____Short attention span
_____Compare symptoms of old age to dementia

_____More than usual difficulty in using technology
_____Difficulty in cooking, or making the same meals each day
_____Not checking mail/mail stacked up
_____Finances in chaos
_____Difficulty driving
_____Getting lost while driving

Doctor Consultation

_____How far along do you think my loved one is?
_____What do I do now?
_____A safety problem to them, family, or anyone?
_____Are any drugs available that may help delay it?
_____Any other health issues?
_____What are the signs of deterioration so that I can explain them to my family and friends?
_____Where can I get more information?
_____What would caregiving entail now and in the future?
_____What signs should I look for in their deterioration where it is best that they are placed in a caregiving facility?

Financial

_____Update will
_____Update assets management information
_____Assign an executor to their estate
_____Determine whether a cremation or a standard burial is appropriate
_____Set up funeral services
_____Streamline finances so it is less complicated for the family
_____Set up Durable Power of Attorney to take effect when mental capacity is limited
_____Determine costs of the home caregiver
_____Determine costs of an assisted-living or other facility
_____Decide what type of care you want your loved one to have

Family Consultation

_____Schedule family members for a meeting
_____Explain as calmly as you can in an organized fashion what is taking place
_____Explain the action you have taken and will take — best to practice your "speech"
_____Ask if there are any questions, and/or disagreements
_____If so, discuss them as calmly as you can
_____Recommend that your family members see your loved one often while there is a memory of who the family is

What to do with your loved one

____Become the 24/7 caregiver?
____Hire an in-home caregiver?
____Move loved one to a care facility?
____Decide what facility may be best for your loved one?

Get loved one's personal affairs in order

_____Contact family attorney or hire one
_____Get a will made or one updated
_____Establish a Durable Power-of-Attorney (for healthcare and finances)
_____Analyze finances and take action as appropriate

Type of Facility

_____Informal
_____In-home caregiver
_____Adult Family/Foster home
_____Assisted living
_____Nursing home
_____Hospice

What care facility is best based on

_____Finances
_____Availability
_____Residence
_____Meals
_____Medicines required
_____Entertainment
_____Transportation

Facility determination

_____How far is it from the family
_____Does your loved one enjoys solitude or crowds
_____Private apartment or shared
_____Ample common space inside and out
_____Can they lock the doors
_____Costs
_____Review last state survey
_____Policy on Medicaid
_____How often is a nurse on site
_____Reasons why loved one may need to move out
_____Who owns the company
_____Any public image of company, incidents
_____Compare all this with family expectations

Facility identified and reviews

_____Admission agreement
_____Resident rights
_____Disclosure of services
_____Pharmacy agreement
_____Photo release
_____Resident handbook
_____Arbitration agreement

Policies and procedures

_____CPR
_____Smoking
_____Cannabis
_____Assisted suicide
_____Pets
_____Outside care
_____Grievances
_____Medicaid policy

Staff observation and evaluation

_____Gift for gab
_____Flexibility
_____Creativity
_____Personal resiliency
_____Reading body language
_____Super-sleuthing
_____Positive, fun attitude
_____Staff turnover rate, why if high
_____Staff training
_____Staffing model
_____Staff—resident integration
_____Staff—staff interaction
_____Engagement
_____Residents' interactions

Residents observation-evaluation

_____Taking items from others
_____Wandering into others' rooms
_____Cussing or speaking gibberish
_____Other unusual actions

When your loved one moves into a facility

_____Visit loved one daily or as often as possible
_____Keep a positive and happy attitude
_____Never yell or even raise your voice in anger
_____Never contradict the loved one
_____Provide entertainment as much as possible, e.g. videos, photos of trips, family, etc.
_____Eat some meals with your loved one
_____Visit at various hours so can also observe staff—patient interactions

Hospice decision and death

_____Get advice from their nurse or doctor
_____Make a decision on hospice
_____Prepare for your loved one's death, e.g. mortuary, etc.
_____Initiate plans for when your loved one is deceased

To hospice or other facility for counseling/discussion group

_____Decide on counseling for you and all of the family
_____Decide whether or not to join discussion group
_____Decide whether or not to volunteer to help others
_____Decide to give to dementia—Alzheimer's-related charities

APPENDIX 2

STORIES OF DEMENTIA— ALZHEIMER'S SUFFERERS

There by the grace of God go I [61]
— John Bradford

INTRODUCTION

The following stories are true and based on the interactions that the authors have had with the people who have Alzheimer's or other forms of dementia; as well as observing the interactions of others. Of course the names were changed and some information masked to protect the privacy of everyone including their families and friends.

These stories are also provided to help you to understand the mindset of the person who has the illness; as well as a possible similar mindset of your loved one or what you may expect from them in the future.

As you read them, remember that you should have compassion, patience, and love for these victims if for no other reason that you may be like them and will want caregivers and others to show you the same kindness, patience, compassion and love.

---

[61] Various sources attribute this quote to Bradford, e.g. see Wikipedia

The Grumpy Resident

There was a male resident, Brian, who lived in the same assisted living facility as Dr. Kovacich's mother. Brian had dementia and whenever he saw Dr. Kovacich he would yell things like, "I don't like you! Get out of here!". The first several times, Dr Kovacich ignored him.

Then one day when the resident said such things, Dr. Kovacich smiled and asked why he wasn't liked. The resident just repeated the same thing. Dr. Kovacich got some coffee and asked the resident to sit down. Surprisingly he did. Then the resident was asked what he liked about his life, etc. At first, the "grumpy" resident didn't say anything.

Dr. Kovacich told Brian he had been in the military. This sparked something in Brian, and he began reminiscing of days gone by. They laughed and visited. They partied as friends and Brian smiled when he saw Dr. Kovacich had returned. They visited and they shared coffee and stories until the day that Brian died.

The Sisters Who Weren't

Kelly and Annie both live in a memory care unit of an assisted-living facility. They both moved in around the same timeframe and were originally introduced to each other during their first meal together. Now, three years later, they are inseparable. One, dressed in her pajamas, will walk to the other's room in the morning and they will get dressed together. They walk hand-in-hand almost the entire day, and chat as if they are long-lost friends. In fact, if you talk with either of them (or both, since they are always together) they truly believe that they are best friends from high school, even though they grew up on opposite sides of the state.

The Non-Family Member Relative

William, a resident living in the memory care unit where my (Dr. Anensen-McNealley) grandfather lives, believes that I am his next-door neighbor. Whenever I come visit Grandpa, William comes up to me, shakes my hand, calls me Ellen (that is not my name), and asks if the grass is growing in the backyard. Over the years I have come to learn that Ellen and Roger's backyards were separated by a wooden fence, and William was obsessed with maintaining a perfectly mowed lawn. Rather than telling William who I really am, I talk with him about the growing grass, or comment on how beautiful his just-mowed lawn is looking. By praising him on his lawn, I acknowledge his passion and serve as a friend, even though none of it is real.

Keep Door Locked Or…

One woman in an assisted-living facility walked next door into her neighbor's room while he was gone. She was found sitting in his chair urinating. She thought she had walked into her bathroom. Another good reason for residents to keep their doors locked when not in the room.

Sometimes Better to Lie

One resident was sitting with his son and having a general conversation. The dad said that his son looked tired and should go to his room to rest. The son told his dad that he did not live there. His dad froze and was confused. We were all taught not to lie, especially to our parents. However, there are circumstances where the lie is sometimes better than the truth when having a conversation to someone with Alzheimer's.

A Need for Compassion, Patience and Love

A son was visiting his mother in the common area of an assisted-living facility. Suddenly the son began yelling at his mother, saying she should understand such-and-such. "Why are you acting so dumb?" he asked her. Can you imagine how his mother felt as she struggled to remember, trying to return to normal with tears in her eyes?

"Selective" Memory

Joan, a daughter-in-law and friend of 37 years, visited Martha, her mother-in-law. Martha didn't know who she was, so Joan introduced herself as Martha's friend. Oddly, Martha still knew her son (Joan's husband)

and recognized him until the day Martha died. Sometimes, long-term memory versus short-term memory is unique to each individual living with Alzheimer's.

Short-Term Memory Loss

A man was sitting with his mom in an assisted-living facility during a holiday gathering. The son left his mom and went to say hello to a woman and her children on the other side of the room. His mom knew the woman for about three years as one of the staff. When he returned, his mom told him to go and stay with his family — meaning the staff member and her children. The son, based on experience and advice from the staff did not say it was not his family. He said that he would later but now wanted to spend time with her. Having Alzheimer's, she soon forgot about that conversation. Another example of short-term memory loss.

Long-term Memory vs. Short-Term Memory

Someone with Alzheimer's could give a detailed description of their wedding day in 1943 but cannot remember what they ate for lunch less than 30 minutes ago.

Assisted-Living Facilities for the Elderly, or People without Dementia

Sometimes, elderly people are tired of living alone, taking care of their home, themselves, cooking, cleaning, and the normal, everyday activities of living. They choose instead to be a resident of an assisted-living facility where cleaning and cooking are done for them and they are amongst residents of their own age. You may come across such individuals where your loved one is a resident.

For residents who do not have a memory loss, living in close proximity to other residents who do have have a form of dementia may be difficult. One woman resident moved to this assisted-living facility after her husband died. She is active in many groups at the assisted-living facility, and has made many friends there. She expresses sadness at one resident with Alzheimer's disease who shares a dining table with her. That woman constantly asks where is her loved one. The woman tells her that she hadn't seen him. The Alzheimer's stricken woman often forgets to eat anything at all and just wanders away from the table right after she sits down.

Such interactions can be depressing to some but may give others the opportunity to be provide support to those stricken with Alzheimer's by just being there for them, talking to them.

Lonely and Repentant

She was a very religious woman who has been repenting, praying for over 50 years for a "sin" she believed she had committed. She usually sat alone and could be seen deep in prayer and thinking of her past. Such people can use additional help; however, they are often seen as just an introverted person who likes being alone. Often, this is not the case but seen that way as others avoid them thinking they want to be alone.

Perceived as a Possible Threat

He was perceived as a possible threat by the staff; however, he had some "mental issues" of a non-dementia nature. It seemed to come over him rather quickly where he became fearful of leaving his room. He eventually moved to an assisted-living facility where he was overly observant of what was going on around him but mostly a "loner" since he was so quiet. He was not diagnosed with any dementia-related disease, but just wanted to live in a place that was quiet and where he did not have to deal with everyday issues. When a volunteer caregiver observed him over time, the caregiver decided to introduce him to other residents, and at first the person was nervous and didn't appear to appreciate the company. However, over time, he opened up, explained why he decided to be a resident of the facility, his life and what he had gone through to cause him to act as he did. He gradually smiled on occasion and opened up to others.

Longing for International Travel

One resident at a facility was there as apparently convinced to do so by a relative. He was an international traveller and had some physical issues that now limited his movements. He loved to talk about his travels and looked forward to the day when he would once again be able to travel to distant lands. One could see in his expression his sadness for being limited in movement. He was a joy to talk to and his face lit up when describing his adventures.

A caregiver sympathized with his plight and said maybe one day you can travel again. His family member heard the conversation and became angry and didn't want the caregiver to encourage him. However, the caregiver knew he would never travel again but wanted to bring some joy in his life, the joy and hope of once again fulfilling his dream. An example of a family member who wanted to him to live in his "harsh, real world". So much for compassion. He died several months later.

Substitute Son

In one assisted-living facility, a resident sat down with a relative of a resident who was having coffee and chatting with other residents. She called him Jim, but that was not his name. The told her he was not Jim but Robert not knowing he violated the "rule" of don't contradict the Alzheimer's patient. She looked confused. She repeated that several times and even a resident told her that he was not Jim. She left.

Later, when her son came to visit, the "substitute son" was also there and explained what had happened and got permission from her son to "impersonate" her son when he was not there. The substitute son continued to do so and she and the relative had many of wonderful conversations about her life, their lives.

She was Born in the Wilderness

One resident loved to reminisce about her life living in the "old days" on a farm in the wilderness where she, through natural birth, had six children. She talked about the difficulties of those days snd how she came to live in the facility. Her face would "light-up" during those conversations snd she often got lost in her history. She enjoyed discussing her life. Although she had Alzheimer's in a more than early stage, her memories were quite vivid. When not having those discussions, she would often be lost in that world; not even recalling eating a few moments ago.

Baseball Lover

Eileen was a 90-year-old woman with Alzheimer's disease living in a small apartment on her daughter's property. Eileen's daughter visited every morning and evening, ensuring her mom had enough to eat and was dressed. Eileen spent most of her days watching TV or playing cribbage with the neighbor. As Eileen's dementia advanced, she became obsessed with baseball; she'd always been a dedicated Seattle Mariners fan and watched their games on TV. Eileen's daughter bought her packages of baseball cards. Eileen began spending hour after hour organizing the baseball cards, putting them in binders then rearranging them later as if they were all new again. Likewise, schedules for all the games were posted on Eileen's refrigerator and her daughter would be sure to call her mom during the day and instruct her on how to get the TV to the right channel so that Eileen could watch the games.

Shortly after baseball season was over, Eileen became withdrawn. She seemed uninterested in the baseball cards, and would call her daughter over and over again at work, telling her that she couldn't find the game on TV. Eileen yelled at her daughter when she'd tell her that the season was over; there was no more baseball to watch on TV. Eileen's daughter hunted for, and eventually found, some old VHS tapes of Mariners baseball games, and began showing those videos (often on repeat) so Eileen could enjoy baseball year-round.

Impending Rising Water

David was a 92-year-old man living with Lewy Body dementia. He moved to a memory care unit when he was 91. He experienced vivid visual hallucinations including seeing dogs in his apartment and kids playing outside in the courtyard. One morning a caregiver came in to David's apartment to help him get ready for breakfast. David had on a pair of sweatpants, and insisted the caregiver help him fold the pant legs up to his knees. When asked why, David responded that "the place is flooding out there" (pointing towards the door that entered the facility hallway). His visual hallucination were so clear to him that, once he got to the dining room for breakfast (still with his pant legs folded up), he warned all his table-mates that their feet were getting wet. A few nights later, David, still focusing on the impending rising waters in the hallway, snuck to the front desk and removed all the computer equipment. He stored them (safely) on the top of his dresser so they would not get wet. This hallucination and the odd behaviors associated with it, eventually went away as David's dementia advanced.

Prim and Proper No More

Alice is a 96-year-old woman living with Alzheimer's disease. Up until a few months ago, she was living with her daughter and son-in-law. Alice walked away from home in the middle of the night, apparently looking for her [long-dead] husband; as a result, her family moved her to a small assisted-living facility where she could get better care.

Alice had spent most of her adult life as a very active member of the Seventh Day Adventist Church. She volunteered at the school there and worked with other dedicated church members in coordinating annual events and weekly after-church potlucks. She and her husband, Ted, had been married 68 years when he died. She was viewed by her children as "prim and proper;" never had a cuss word eeked from her lips, nor did she speak poorly of anyone.

As Alice's dementia advanced, her language changed. She would be overheard cussing at the caregivers, calling them such names as "stupid little bitch" or "bastard." She flirted with a male caregiver, even grabbing his bottom once when he was helping her get dressed.

Alice's family was mortified; her daughter voiced embarrassment at her mom's behavior. The owner of the assisted-living facility explained to the family that sometimes a person's past character and personality changes as the brain cells die, and to not take these behaviors to heart. "Somewhere, deep inside your mom's mind, she is still there. Remember *that* mom, not the one you're experiencing today."

Acute Confusion or Dementia?

Jules is a 74-year-old woman living across the street from her adult son. She spends a lot of time at home, alone, sewing and watching TV. She comes to her son's house a few evenings a week for dinner, but otherwise she lives her own life. She has been diagnosed with chronic pain and insomnia, and had a stroke years ago, from which she's recovered. She manages her own medications and is very reluctant to have any of her kids know what medications she takes.

Early one morning Jules shows up on the front porch of her son's house; she is dressed in a long dress and high heels. When her son asks her what she's doing, she insists that it's time for a big family dinner at a local restaurant and "we're running late – you'd better get dressed so we can make it there on time."

Jules' son is deeply concerned about his mom's new confusion and takes her to the emergency room, where the doctor discovers that Jules has delirium (acute onset of confusion), most likely caused by mixing prescription medications. Jules admits that she's been taking more of the sleeping pill than her doctor prescribed, because it wasn't working, and had also increased her pain medication in the hopes that the two combined would give her some relief from sleepless nights. To make matters worse, Jules had also taken some over-the-counter Benadryl because she knew that a side effect of this medication is drowsiness.

Sudden onset of confusion can mean delirium, not Alzheimer's. It is imperative to have your loved one checked out immediately for this condition – it could be an infection, medication interaction or allergy, dehydration, fever, or a thyroid disorder.

SUMMARY

Alzheimer's and other dementias affect the persons in different ways, and impact families on several different levels. Whether your loved one lives at home or in a facility, it's likely that you will encounter situations where your creativity is key to keeping the peace or engaging the person in life. This "go with the flow" attitude is one not only beneficial to professional caregivers, but to family members as well. By keeping the thought that the person with Alzheimer's is always right, the relationship you have with your loved one will continue strong, albeit likely to be very different when compared with before Alzheimer's was a part of their life.

If you are visiting a loved one in a caregiving facility, and you see a resident sitting alone, take the time to introduce yourself and your love one. Strike up a conversation. They have such wonderful and amazing stories to tell about their lives. You'll be glad that you did.

If you are visiting a loved one in a caregiving facility, and you see a resident sitting alone, take the time to introduce yourself and your love one. Strike up a conversation. They have such wonderful and amazing stories to tell about their lives. You'll be glad you did.

APPENDIX 3

MEMORIAL TO HELEN M. KOVACICH

This appendix is written as a memorial dedicated to my mother.

Then She Was Gone

On August 14, 2014, mom died of natural causes. She was 93 years, seven months young. She was cremated in Mt Vernon, Washington, and her ashes buried next to my father in Las Vegas, Nevada.

She has taught me so much about compassion, patience, love and charity in her last years of life. If for no other reason than that, I will be forever grateful as she has made me a better human being.

As I figure it, there are three possibilities for her after death…

- If there is nothing waiting for us but being worm food, well, at least my parents are still physically together.
- If there is no Heaven or Nirvana, or at least not now for her, but there is reincarnation, I hope that her spirit once again is able to find that of my father's and they can have another great life together. If reincarnated without my father, hopefully she will have a life with great adventures with another great partner.
- She had waited since my father's death in 1977 to join him. If there is a Heaven, I believe she has been more than good enough to go there. I hope they are enjoying that eternity together along with her other relatives and friends. What a homecoming they could be having.[62]

---

[62]

> *Mom's one and only tattoo gotten at age 92, because she lost a bet. She later wanted a lady bug on the other arm but her skin was too old to take the ink as noted by blue "ink bleeding" from her blue butterfly tattoo on her forearm. She was the oldest person Mike Bandana of Bandana Arts in Anacortes, WA, ever tattooed in more than 30 years of tattooing clients.*

During the 10-plus years my mother lived nearby, I spent a great deal of time with her. So much so that I never saw the changes brought on by Alzheimer's, until 2010. During the more than three years I spent visiting her, almost daily for hours in her assisted living facility, talking to the staff and other residents, I found that I was quite ignorant of this disease.

I decided to learn all I could through research and observation of my mother's decline, interviews of caregivers, and spending time with other people with Alzheimer's.

I decided that after my mother's death, I would write a book about what family members need to know and do if their loved one was diagnosed with Alzheimer's so they wouldn't be as ignorant as I had been about this disease.

I met my co-author at my mother's assisted-living facility. We agreed such a book would be helpful and decided to write it together, all profits going to an Alzheimer's charity.

Eulogy for Helen M. Kovacich

When my mother died, I celebrated her death. Why? Because I was happy that her suffering was over. No more bad back. No more bad knees. No more Alzheimer's, throwing up, seizures, heart stoppages. No, that is now all gone. But is she at peace? Absolutely not! My mom is not one to sit around and I am sure she is not doing that now either, even if it is Heaven. It's just not in her nature.

Surviving the death of her beloved husband and my father in February 1977, being on her own for so many years for the first time in her life was difficult. However, she surprised me as she turned out to be a tough woman.

My dad only taught her to drive at about age 57. He was old fashioned and by that I don't mean she was subservient. Just the opposite. He worshiped her and did everything he could to make her happy and carefree. So, it was surprising how she was able to take care of herself from the very beginning — following his death.

My mom often wondered why she was living so long when she was the last of her clan. I told her that it was because she was staying on so I could learn compassion, patience, love and charity from her and when I had learned that, she would die. Well, mom I guess I have learned.

Some thought my mom was a pain in the ass and yes, sometimes, she was, like sometimes we all are. My mom was German—Hungarian and married a Croatian. Now that is a formula for enjoying life — a combination of beer, wine, good food, family and good friends, while celebrating just being alive. She always said to give her a "shot and a beer" but she didn't drink. Our clan was known for being a little louder than most, and we tend to express our joy of life as much as we can — loudly.

So, my mom, and I being my mother's son, while at Harbor Towers sometimes got stern looks from some of you when we talked too loud, laughed too loud, and using what some say is "profanity". Mom was fond of saying, "hell, damn, shit" isn't swearing. Only one word was considered by her to be out-of-bounds and of course that was the big "F" word. She used to say she hated that word but she never had a good explanation why that word was so terrible.

My mother had a couple of sayings that she loved to repeat. One was, "Of all the things I miss, I miss my mind the most". Then she used to laugh. However, the laughing at that saying gradually stopped beginning in the summer of 2011 when she was officially diagnosed with Alzheimer's and she gradually did start missing her mind.

The other saying she was fond of was when she talked of still being alive at her age and the last of her clan: "I guess God doesn't want me up there causing trouble". And then she would laugh. Well, it looks like even God couldn't keep her out of Heaven any longer.

So now, she's up there with her husband (my dad), friends and relatives. Up there in Heaven, raising hell and I'm sure God and his angels are rolling their eyes and wondering if they could get her soul reincarnated and back down to earth to delay her eternity in Heaven for just a little while longer. They probably feel she has already been up there an eternity.

Then again, maybe God loves her company. After all, he's got to have a sense of humor, he made us didn't he? Give'em hell mom and keep 'em laughing.

APPENDIX 4

THOUGHTS OF GRANDMA AND GRANDPA FILMORE

My maternal grandparents raised their children in a rural area in Washington State. They remained in their home well beyond their physical abilities: in their mid-80s they could no longer climb the stairs to their bedroom. My parents built them a small home only steps from their own, where the family could provide whatever care might come as they continued to age.

While physical ailments popped up here and there – a broken hip for Grandpa, pneumonia for Grandma – the real robber was memory loss. Grandma's came first; Alzheimer's snuck into the household with snide remarks and slamming doors. She became obsessed with her small dog while completely disregarding Grandpa, the house, eating, and bathing. She layered her clothes, dirty ones piled on, with one clean layer on top. Her adamant refusal for any assistance left her in the home, without needed help, until her death.

But before that, she became cruel. Typical of Alzheimer's, Grandma verbally scrutinized my mom – her only daughter – for nothing more than existing. My mom was Grandma's primary source of help, however discreet, yet Grandma's disease did not allow her acceptance and grace but rather ridicule and outright hatred. My mom took the brunt of Grandma's memory loss and lost their mother—daughter relationship along the way.

I remember one night when I was staying the night at my grandparents' house, caring for Grandpa after hip surgery. It was cold outside and in the middle of the night I took Grandma's dog outside for a quick potty break.

Grandma came to the door in her filthy, lace-torn nightgown and 1950's hair curlers, sneered at me, and yelled, "Ah-ha! Finally, I got you outside where you belong! You are NOT welcome back." She slammed the door and locked it behind her. Thankfully I had the key, and let myself (and the dog) back in once Grandma returned to her bedroom.

Despite the meanness, Grandma's Alzheimer's was "run of the mill." Typical accusations that we'd overfed her dog and that the car's battery was dead (my dad had dismantled the car's engine so Grandma could no longer drive). But it was my Grandpa's memory loss that hit me harder.

Grandpa willingly went to an assisted-living when he needed more care than we could provide. Grandpa's memory was already bad, but it became progressively worse until it far surpassed any physical ailments limiting his functioning. He did not have Alzheimer's like Grandma, though. His dementia was called Lewy Body dementia. This form comes with so many more unique characteristics that warranted special attention and care.

Grandpa experienced visual hallucinations. He saw dogs in his assisted-living apartment. Those dogs were so real to him that he would worry that they would bite his grandkids or we would inadvertently step in their waste (which apparently he could see on the floor). He had nightmares and his body acted them out as he slept; he often fell out of bed because of them. He believed that the assisted-living's hallways were flooding; he would even roll his pants legs up before going to the dining room for meals. Any medications the doctor tried to manage his behaviors only made him worse – either paranoid beyond comprehension or sacked out and sleeping the majority of the day and night.

One day he stopped eating and drinking. His body spiked a fever without any associated infection (this is also part of this type of dementia). It's as if he knew that he could be in control, finally after years of suffering with this dreaded disease. He died under hospice care, with his family at his side, one day before his blessed Seahawks lost the Super Bowl. Thank goodness he did not have to see that game.

APPENDIX 4

ABOUT THE AUTHORS

Dr. Gerald L. Kovacich
Gerald and Helen on Mother's Day, 2014, the year of her death.

Dr. Kovacich has spent more than 50 years (so far) as a global traveler, poet, author and observer of Life, as he tries to follow the philosophy of living a life of compassion, patience, charity and love for all life forms.

He is the author of more than 20 books, both fiction and non-fiction; several translated into Russian, Chinese and Japanese.

His travels have taken him to many parts of the world where he has learned their cultures, beliefs, languages and enjoyed their stories of Life, including death rituals. However, nothing had prepared him for the day the doctor diagnosed his mother with Alzheimer's.

Dr. Vicki L. Anensen-McNealley

Vicki is a Washington State girl through and through, having grown up on a farm just outside a small town in Western Washington and promptly returning there after completing nursing school at Washington State University.

Vicki has been involved in assisted-living facilities in a number of different capacities since 1999. Firstly, as a state licensor/investigator and a quality improvement consultant, then as a curriculum developer for dementia training, and currently as a corporate nurse for an assisted-living company. She also moonlights in a memory care unit on the weekends, overseeing caregivers and providing nursing care for residents.

During her Master's in the nursing program at the University of Washington, Vicki got the curriculum development bug – she has created numerous training programs, many focused on the care and rights of Alzheimer's patients, for nurses and caregivers alike. She has spoken across the nation on the unique needs of people living with Alzheimer's disease, with much emphasis on quality of life, sexuality, and the individual's happiness. She also provides dementia training for caregivers and nurses working in long-term care in Washington State.

Both authors have years of hands-on, personal experience in dealing with this disease; as well as other forms of dementia. Their first-hand knowledge is based on dealing with and caring for their loved ones with dementia. Dr. Anensen-McNealley has had decades of experience as a PhD, registered nurse in a leadership role in dealing with dementia and specifically Alzheimer's.

www.ingramcontent.com/pod-product-compliance
Lightning Source LLC
Chambersburg PA
CBHW020658220526
45464CB00001B/490